T0340714

Statistical Robust Design

Statistical Robust Design

An Industrial Perspective

Magnus Arnér

Tetra Pak Packaging Solutions, Sweden

WILEY

This edition first published 2014
© 2014 John Wiley & Sons, Ltd

Registered office
John Wiley & Sons Ltd, The Atrium, Southern Gate, Chichester, West Sussex, PO19 8SQ, United Kingdom

For details of our global editorial offices, for customer services and for information about how to apply for permission to reuse the copyright material in this book please see our website at www.wiley.com.

Library of Congress Cataloging-in-Publication Data

Arnér, Magnus, author.
  Statistical robust design : an industrial perspective / Magnus Arnér.
    pages cm
  Includes bibliographical references and index.
  ISBN 978-1-118-62503-3 (cloth)
  1. Industrial design–Statistical methods.   2. Robust statistics.   I. Title.
  TS171.9.A76 2014
  745.2–dc23

                                                    2013046030

A catalogue record for this book is available from the British Library.

ISBN: 978-1-118-62503-3

Set in 10/12pt Times by Aptara Inc., New Delhi, India
Printed and bound in Singapore by Markono Print Media Pte Ltd

1   2014

Contents

Preface

For several years I have been waiting for a book to be published about robust design that I really like. The problem is not that books about robust design are nonexistent. On the contrary, there are many books on the topic, and some of them are really good. But I did not feel that anyone really was speaking directly to me. Slowly, really slowly, an insight was growing within me that the number of people who have a good statistical background and have been working as industrial practitioners with robust design is fairly small. Since it is within this small group that an author for the type of book I am missing is most likely to be found, there are not an enormous amount of potential writers. In addition, industrial practitioners are not as keen on writing books as university teachers. I came to ask myself if I was the one to write such a book. The result is the product you are holding in your hand.

There are many ways of viewing robust design. It is, for example, a methodology in product development. Thus, robust design can be included as part of a book about product development. Robust design is also quality engineering (if that is seen as something separate from product development). In the same way robust design can be part of a book about product development, it can be part of a quality engineering book. A third view takes off from the way in which knowledge that can be used to accomplish robustness is gathered, namely design of experiments. This book is not intended to be a book about product development or quality engineering in general, a book about design of experiments. It is intended to be a book about robust design. Neither product development in general nor the basics of design of experiments with two-level factorial designs are covered in the book. A potential reader may then be an engineer with some experience of product development and some basic knowledge in design of experiments. I have personally used the first six chapters of the book as a part of the Design for Six Sigma training for exactly this type of audience. However, even if this may be the primary target group, the book can hopefully be useful to a much wider audience, including engineering and statistics students.

Robust design may have been used for a very long time but without the terminology and structure it has today. In its present form it emerged in Japan after the Second World War and reached the United States around 1980. Thus it has some history. During this time there has been a constant stream of research papers. Several among them were written decades ago. However, from the point of view of industrial practice it is first and foremost during the last 15 years that robust design has become a natural element in product and process development. Earlier on, some companies were taking robust design seriously, but only a few. Nowadays it is hard to find any major company not working with robust design, or at least not claiming they do. There are a number of reasons for this increased industrial interest. One is Design for Six Sigma (DFSS), in which robust design is a major element. With an ever increasing number of engineers educated in DFSS the companies are growing their competencies and capabilities in the robust design field. The usage of design of experiments has also increased substantially during the last decade, and since this is such a central element in robust design it has increased the ability to design for robustness. Thus, the experience and knowledge among engineers has reached a critical level. Another reason is the availability of dedicated software, especially the software that can be integrated in the software architecture already in place in the industry, such as robust design modules in Computer Aided Design (CAD) software and even more importantly finite element solvers.

I do not know when I first came in touch with robust design. I remember that I wrote a review of a book on the topic in 1992, so it must have been before that. However, it was not until 2002 when I was working for an automotive company that I really had the opportunity to apply it at work. During that time I had the opportunity to learn from Shin Taguchi (the son of Genichi Taguchi who had founded modern robust design half a century earlier) whose company was engaged in some work at the automotive company during that time. Since then robust design has been a major part of my work, with applications in both the automotive and packaging industries.

Example 2.1 is based on work by Mats Martinsson, Example 3.6 and Example 5.4 on work by Markus Florentzson, and Exercise 5.2 on work by Ulrika Linné, Hossein Sohrabi, and Andreas Åberg. Besides them, there are several people who have been helpful and encouraged me in the work on this book. I would like to mention Bo Bergman, Johan Olsson, and Pietro Tarantino.

This book contains an accompanying website. Please visit www.wiley.com/go/robust

Magnus Arnér
Lund, Sweden

1

What is robust design?

1.1 The importance of small variation

When mass production started in the dawn of the industrial revolution, variation came in the focal point of interest. An early example that illustrates this concerns mass production of guns. The French gunsmith Honoré le Blanc realized the importance for guns to have interchangeable parts. His solution was the invention of a system for making gun parts in a standardized way. The problem that challenged le Blanc is the same as in any modern day manufacturing, as, for example, in the production of bolts and nuts. It shall be possible to pick a bolt and a nut at random that fit together. This requires that the variation in diameter, in roundness, and in thread pitch is small from bolt to bolt and from nut to nut. Unless this is the case, there will be a substantial amount of scrapping, or even worse bolts that crack or fall off while they are in use.

Before the industrial revolution, this problem was handled by good craftsmen. In the industrial era, this was not an option anymore. The importance of managing the variation became obvious. Several approaches emerged. Specifying the tolerance limits was one of them and even if the gunsmith le Blanc did not get many immediate followers in France, some Americans saw the potential of his ideas and implemented them at the armoury in Springfield. This is sometimes considered as the birth of tolerance limits (which is not quite true as tolerance limits are much older than this).

To quote Edward Deming, a forefront figure in quality engineering, 'Variation is the enemy of quality.' The bolt and the nut is one example. Another one is thickness variation of the plastic film on the inside of a milk package–a plastic film preventing the beverage from coming in contact with the aluminium foil that is present in most milk (and juice) packages. If this thickness varies too much, it may occasionally happen that there is a point with direct contact between the beverage and the aluminium foil. However, it is not the fact that there is a contact point that should be the centre

Statistical Robust Design: An Industrial Perspective, First Edition. Magnus Arnér.
© 2014 John Wiley & Sons, Ltd. Published 2014 by John Wiley & Sons, Ltd.
Companion website: www.wiley.com/go/robust

of interest. The focus should rather be the size of the film thickness variation. The contact point is just a symptom of this problem.

Investigations show that a substantial part of all failures observed on products in general are caused by variation. With this in mind it is obvious that variation needs to be addressed and reduced. The issue is just how.

This book is about random variation, or more specifically how to reduce random variation in a response variable y, but not just any way to reduce this variation. It will not be about tightening tolerances, not about feedback control systems, and not sorting units outside the tolerance limits. The focus is solely on preventing variation to propagate. It is this approach to variation that is called robust design.

1.2   Variance reduction

Example 1.1   Consider a bar that is attached to a wall. There is a support to the bar and a random variation in the insertion point in the wall, as sketched in Figure 1.1. We are interested in the position (x, y) of the end point of the bar and that its variation is small.

The variation of the end point position can be reduced in two fundamentally different ways. One is to reduce the variation of the insertion point in the wall. It may be costly. Typically it can involve investment in new and better equipment. However, it might be another way to reduce the variation of the end point position, namely to move the support. In that way the design becomes less sensitive to the variation in the insertion point (Figure 1.2). This is what we mean by robust design: to make the variation of the output insensitive (robust) against incoming variation.

Example 1.2   Two metal sheets are attached to each other. There are two holes in each one and they are attached to each other using two bolts (Figure 1.3). Assume that the maximum stress σ_{max} in the metal sheets is of interest to us. A small variation in the position of the holes, or rather distance between them, affects this stress.

Suppose that this stress should be minimized. This can be achieved by increasing the precision of the positions of the holes. It can also be achieved by changing the

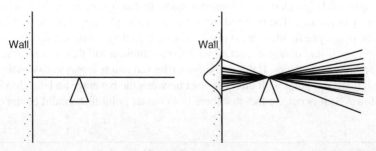

Figure 1.1   A bar is attached to a wall. There is a support to keep it in place. If there is a variation in the insertion point, there will be a variation in the end point.

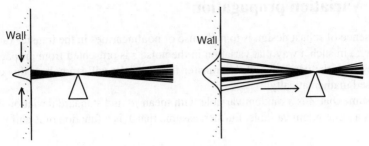

Figure 1.2 A bar is attached to a wall. The variation in the end point can be reduced in two fundamentally different ways, namely by reducing the incoming variation or by making the design robust against this variation.

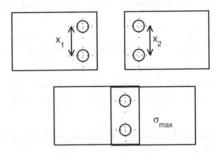

Figure 1.3 Two metal sheets are mounted together. Since there is a variation in the attachment position, there will be a variation in the maximum stress.

design so that one hole is exchanged for a slot (Figure 1.4). In that way, the variation in the response, the maximum stress, is decreased without reducing the variation in the sources of variation, the hole positions. The stress is robust to the hole position.

We have seen two examples of robust design, the end point of the bar and the stress of the metal sheets. For both of them, ways to reduce the variation of the output without reducing or removing the original source of variation were pointed out.

In robust design, the original source of variation is called noise. This noise is typically sources of variation that the engineer cannot remove or even reduce, or something that can be reduced but at a considerable cost.

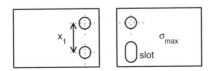

Figure 1.4 Exchanging one hole for a slot will make the stress robust against the variation in the hole position.

1.3   Variation propagation

The essence of robust design is to make use of nonlinearities in the transfer function $y = g(x, z)$ in such a way that variation in the noise z is prevented from propagating. The key is in the derivative of the transfer function. We will study how this can be expressed mathematically.

Assume that Z is a random variable with mean μ_z and standard deviation σ_z and that x is a nonrandom variable. Further, assume that Y is a function of Z and x,

$$Y = g(x, Z).$$

For example, Z can be the distance from the nominal attachment point and x the distance from the wall of the support in Example 1.1. Since Y is a function of the random variable Z, it is also a random variable. Taylor's formula gives

$$g(x, z) = g(x, \mu_z) + g'(x, \mu_z)(z - \mu_z) + \frac{g''(x, \mu_z)}{2!}(z - \mu_z)^2 + \cdots$$

where all the derivatives are taken with respect to z. This will be applied to study how the mean and variance of Z propagate to Y. For the mean of Y we obtain

$$E[Y] = E[g(x, Z)] \approx g(x, \mu_z) + g'(x, \mu_z)E[Z - \mu_z] = g(x, \mu_z)$$

and for the variance

$$Var[Y] = Var[g(x, Z)] \approx Var\left[g'(x, \mu_z)(Z - \mu_z)\right]$$
$$= \left(g'(x, \mu_z)\right)^2 \sigma_z^2. \tag{1.1}$$

These are called Gauss' approximation formulas.

Let us apply this to Example 1.1. Define

$$Y = \text{height of end point}$$
$$Z = \text{insertion point in wall}$$
$$x = \text{position of support, distance from wall}$$

and

$$L = \text{length of bar.}$$

Note that x has a different meaning here than in Example 1.1. We obtain

$$Y = g(x, Z) = Z + L \sin\left(\arctan\left(-\frac{Z}{x}\right)\right)$$

and

$$Var[Y] \approx \left(g'(\mu_z)\right)^2 \sigma_z^2 = \left(1 - L\left(\frac{x \cos\left(\arctan\left(-\frac{\mu_z}{x}\right)\right)}{x^2 + \mu_z^2}\right)\right)^2 \sigma_z^2. \qquad (1.2)$$

In robust design, Z is called a noise factor or variable that the engineer cannot fully control and x a control factor since it can be controlled.

The main difference between classical ways to approach variation and robust design is which factor in Gauss' approximation formula (Equation 1.1) to focus on.

Gauss' approximation formulas:

$$\mu_y \approx g(\mu_z)$$

$$\sigma_y^2 \approx \left(g'(\mu_z)\right)^2 \sigma_z^2$$

It is how the second one of these, $\sigma_y^2 \approx (g'(\mu_z))^2 \sigma_z^2$ (see Figure 1.5), is used that makes robust design different from traditional ways to reduce variation. In traditional engineering, the variance of the variation source, σ_z^2, is reduced in order to reduce the variation of the response, σ_y^2. In robust design the objective is still to reduce σ_y^2, but it is done by reducing $g'(\mu_z)$.

1.4 Discussion

There have always been engineers that have made their designs insensitive to factors that are outside their own control. However, the modern development started in the 1950s, when Genichi Taguchi divided the factors in designed experiments into two different categories, noise and control factors. For some decades, it was primarily in Japan that these ideas attracted any attention.

Around 1980 the ideas of Taguchi reached North America and Europe. The basic principles, like dividing the factors into two different categories, were highly appreciated but some other ideas of Taguchi were more controversial. These controversies will be touched upon at several occasions in this book.

Even if robust design has been around for quite a long time, it is only since around the year 2000 that the application of it in industry has started to grow. One reason is that many companies have started programmes for Design for Six Sigma, DFSS, where robust design plays a central role. Another reason for the growth is the availability of software for CAE (computer aided engineering) based robust design.

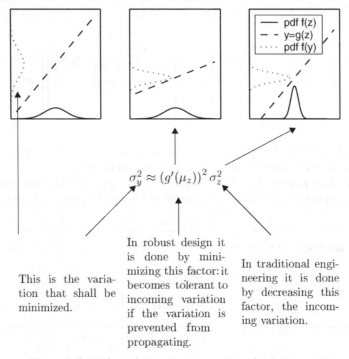

$$\sigma_y^2 \approx \left(g'(\mu_z)\right)^2 \sigma_z^2$$

This is the variation that shall be minimized.

In robust design it is done by minimizing this factor: it becomes tolerant to incoming variation if the variation is prevented from propagating.

In traditional engineering it is done by decreasing this factor, the incoming variation.

Figure 1.5   Robust design is one way to reduce variation.

1.4.1   Limitations

Since robust design is a broad field, we need to limit ourselves in this book. Two of these limitations are worth mentioning. When Taguchi puts robust design in a framework, he mentions three steps:

(i) system design;

(ii) parameter design;

(iii) tolerance design.

The first one of them, system design, deals with the development of the concept. It is about developing a concept that in itself is robust. The second one, parameter design, is a particular kind of design of experiments (DOE). It deals with the question of how the factor levels (parameter levels) shall be selected given a certain concept. The third one, tolerance design, is left aside in this discussion.

The focus of this book is on the second step, parameter design. It might be somewhat unfortunate, since it is often in the development of the concept or the selection of the concept that robustness is born. However, system design is often difficult and hard to grasp. There is no data available at that point of time and limited possibilities for data collection, so the work has a tendency to become more of a speculation than data driven analysis. There are of course exceptions, often based on a solid understanding of the mathematics and physics of the concepts and intensive usage of the variation propagation formulas of Section 1.3. Therefore, robust design in the concept development phase may look beneficial from a theoretical point of view, but when it comes to the practice it mostly turns out that the benefit is limited. Therefore, the focus of this book is on the step that Taguchi calls parameter design.

Even if we limit our discussion to the DOE related part of robust design, parameter design, there are still different ways to perform the experiments. They may be virtual or physical. The main stream in the book is physical experimentation, but one chapter is devoted to the specific kind of issues arising in CAE based robust design.

1.4.2   The outline of this book

The order in which the topics are introduced in this book is not chronological. It does not start off by deciding what to make robust, followed by a discussion on what it should be robust against. It will rather start with a fairly late step, namely the design of experiments and the analysis of the results (Chapter 2). Our reason is to quickly get into a stage where we can do calculations and from this provide an understanding for the preparations needed before these steps are reached. Starting with DOE will make it possible to quickly come to practical exercises.

It is followed by a chapter about noise factors, and not until thereafter will the issue of what to make robust be reached. The attribute to make robust can be a single value like the mechanical tension of Example 1.2 or a relation between a third type of factor, called a signal, and a response. It is in turn followed by a deeper discussion about factors. It is not always obvious whether a factor is a noise factor (that the engineer cannot fully control) or a control factor.

Already in the discussion about response variables, some of the specific ideas of Taguchi are introduced. In Chapter 5, more will come. It covers a special type of DOE, inner and outer arrays, that originates from Taguchi. The chapter includes a discussion about the pros and cons of such designs compared to the ones covered earlier on in the book.

Once the analysis has been performed, a mathematical model is built and the design that is the most robust is pointed out. However, in order to have confidence in this, it must be checked whether the mathematical model is trustworthy and the design pointed out as the most robust one really is good enough.

Up to Chapter 5, the target value is supposed to be of the type 'nominally the best'. Chapter 6 covers other types of target values, such as 'the larger the better' and 'the smaller the better'.

In most texts on robust design, it is stated that noise factors must be possible to control in the experimentation despite the fact that they are not controllable in real life. Unless they are, they cannot be included in the designed experiment and it will be impossible to make the design robust against this noise. This is not quite true. Even if noise factors cannot be controlled in the experimentation, it may still be possible to make the design robust against them. The remedy is regression analysis. However, this requires that noise factors are possible to observe and measure. Regression is the topic of Chapter 7.

Then, a number of chapters will follow where various topics of robust design are covered. These chapters may not be of interest to each and every reader, but will be invaluable for those who need a deeper and wider knowledge in the field. The first one of them (Chapter 8) goes deeper into a concept introduced in Chapter 7, namely various types of optimality criteria for robustness. Chapter 8 will also introduce concepts that pave the way to Chapter 9, which is about the special types of experimental methodology that are useful in virtual experimentation and simulation based robust design.

With mass production, there is no or limited possibility to fine-tune the geometry of each and every produced unit (e.g. by shimming). The geometry of the final product must be robust against the geometry variation of its components and of the manufacturing operations. It turns out that Monte Carlo methods are very useful for robust design applied on variations in geometry. This possibility has caught some attention in the industry, especially in the automotive industry, where the usage is fairly wide spread. These Monte Carlo methods are the topic of Chapter 10.

In the last chapter in the book, the history of robust design, or rather Taguchi methods, is presented and the ideas of Taguchi are discussed.

Exercises

1.1 When two objects of material are attached together with glue, the glue can be oxidized in the moment just before the attachment to increase adhesion. It is done by blowing gas on to the glue. However, the dosage (amount) of this gas cannot always be fully controlled. There is a variation around the nominal value as marked with the Gaussian curves of Figure 1.6. In some applications, there is a saturation point for the gas dosage, as in Figure 1.6. Which nominal setting of the dosage is best from a robustness point of view if the adhesion should be maximized, x_0 or x_1?

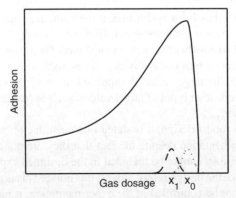

Figure 1.6 The adhesion between the two objects may depend on the gas dosage.

1.2 Consider Example 1.1 where a bar is attached to a wall. Denote the end point position (y_x, y_y), the length of the bar L, the attachment point in the wall Z, and the position of the support (distance from wall) x. For simplicity, assume that $\mu_z = 0$ and that the height of the support always takes value zero.

    (a) Sketch the variance of Y_y as a function of the support position x by using Equation (1.2). Let x be in the range $0.2L$ to $0.95L$, where L is the length of the bar. What happens when $x \to L$?

    (b) Use trigonometry to prove Equation (1.2) and to derive the function $Y_x = g(Z, x)$ for the abscissa coordinate of the end point position.

    (c) Use Gauss' approximation formulas to obtain the mean and variance of Y_x and reflect on the result. Does the result make sense? If not, explain why!

    (d) Use a higher order Taylor expansion to and sketch the variance of Y_x as a function of the support position.

2

DOE for robust design, part 1

2.1 Introduction

Design of experiments (DOE) plays a key role in robust design. It is through DOE that knowledge about how a design can become robust is obtained. However, there are some fundamental differences between traditional DOE and DOE for robust design. One important difference is that there are two different types of factors in DOE for robust design, noise factors and control factors. Noise factors are factors against which the response should be robust. Control factors are factors whose values can be decided by the engineer. Another difference is that in DOE for robust design, the primary interest is a type of interaction, namely the interaction between control and noise factors.

2.1.1 Noise factors

Noise factors are, as mentioned, factors against which the response should be robust. They cannot be controlled in real life, at least not for a reasonable cost, but they can be controlled in the experiment. Typical noise factors are manufacturing variation, customer and operator behaviour, and environmental conditions such as ambient temperature or relative humidity.

Let us take an example. Assume that the perceived image quality of a television screen should be robust against the angle of viewing and whether or not the spectator wears glasses. Then the angle of viewing and glasses/no glasses are noise factors (Figure 2.1).

Statistical Robust Design: An Industrial Perspective, First Edition. Magnus Arnér.
© 2014 John Wiley & Sons, Ltd. Published 2014 by John Wiley & Sons, Ltd.
Companion website: www.wiley.com/go/robust

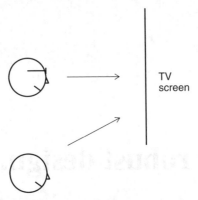

Figure 2.1   If we want the perceived image quality to be the same no matter whether the spectator uses glasses or not and no matter what the angle of viewing is, then glasses/no glasses and angle of viewing are noise factors.

2.1.2   Control factors

Control factors are factors that

- are expected to affect the response;
- can be selected by the engineer to have a certain value.

Control factors are factors about which the engineer can take decisions and that are supposed, or at least suspected, to affect the response. It can typically be geometry parameters, material properties, and parameters or types of algorithms in software codes. For geometry parameters and material properties, it is only their nominal values that are control factors; the random variation around the nominal is noise, so that the actual value taken by, say, the density of some material is the sum of a control factor (the nominal value) and a noise factor.

2.1.3   Control-by-noise interactions

The key to robustness is control-by-noise interactions. The idea is to select the values of the control factors in such a way that the response becomes insensitive to the values taken by the noise factors. Let us take an example. Consider a 2^3 factorial design with two control factors and one noise factor (Table 2.1). In Figure 2.2 the result is illustrated in a factor–effect plot for the control-by-noise interactions. It shows that if control factor $A = -1$, then the response is fairly insensitive against the value the noise factor takes. The response is robust against noise C. Factor B cannot be used to accomplish robustness.

When the noise and control factors are combined into one common factorial array, as in Table 2.1, it is called a 'combined array'. Just the fact that there is a specific name for this indicates that there is another alternative as well. This alternative is

Table 2.1   The outcome of an artificial experiment
with two control and one noise factors.

Control		Noise	Response
A	*B*	*C*	*y*
−1	−1	−1	22.4
1	−1	−1	27.0
−1	1	−1	25.1
1	1	−1	29.7
−1	−1	1	22.3
1	−1	1	19.9
−1	1	1	25.3
1	1	1	22.0

called 'crossed arrays', 'product arrays', or 'inner and outer arrays' and is studied in
Chapter 5. In the present chapter we will stick to combined arrays.

2.2   Combined arrays: An example from the packaging industry

Example 2.1   For a milk package, it is important that the carton board does not
come in direct contact with the milk. For this purpose, the edge of the carton board is
sealed with a plastic strip. For some types of packages, this strip is wrapped around
the edge, as shown in Figure 2.3.

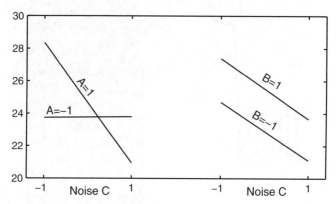

*Figure 2.2   If the control factor A is set to $x_A = -1$, then the response will be robust
against noise C.*

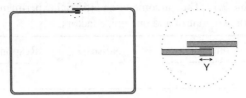

Figure 2.3   Intersection of a milk package with the area of the strip magnified. The plastic strip prevents the carton board from soaking up the beverage. The target value of Y is m.

There was a concern that the variation of the length Y (marked in Figure 2.3) of the strip would be too large, especially for the strip on the side in contact with the milk. A major reason for this variation was supposed to be the temperature of the board while applying the strip. It was therefore decided that a robust design study should be performed with the length Y as the response and the temperature as the noise. The target value of Y is m.

The strip is applied to the packaging material before it is cut into the sheets of the individual packages and while the board is running, as shown in Figure 2.4. There were five control factors in this experiment. They are given, along with their levels, in Table 2.2.

One of the control factors is the strip tension. Its different levels in the experimentation are accomplished by changing the material of the weight of the so-called strip accumulator on the lower left of Figure 2.4. There are two wheels, an upper and a lower (Figure 2.4). They have profiles in order to fold the strip around the edge. The

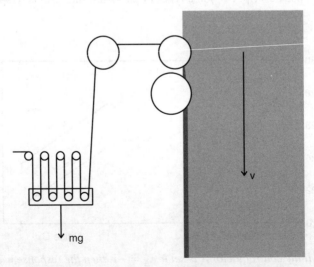

Figure 2.4   A sketch of the strip application.

Table 2.2   The control factors and their levels in the strip application experiment.

Factor	Low level	High level
A: Strip tension	Low (Al weight)	High (steel)
B: Rolling resistance upper wheel	Ball bearing	Plain bearing
C: Upper wheel profile	V shaped	U shaped
D: Lower wheel, property 1	Low	High
E: Lower wheel, property 2	Low	High

four remaining control factors are all properties of these two wheels. The levels of factor C are explained graphically in Figure 2.5. Finally, there is a pair of wheels that are perpendicular to the two first wheels but are not drawn in the figure or included in the experiment. In this pair there is one wheel on each side of the board with the aim to press the folded strip on to the board.

2.2.1   The experimental array

Five control factors and one noise factor sum to a total number of six factors. Since most of these factors are qualitative rather than quantitative, it does not make much sense to include a centre point in the experiment. A two-level full factorial design gives $2^6 = 64$ combinations. This was considered to be too much, so a half fractional with $2^{6-1} = 32$ experimental settings rather than 64 was used. The measurement results are given in Table B.1 in Appendix B.

2.2.2   Factor effect plots

The data analysis can be carried out in many different ways, both analytically and graphically. A factor–effect plot is a good starting point.

We will look at two different plots simultaneously, control-by-noise interactions and control factor main effects (Figure 2.6). Let us analyse these plots to learn

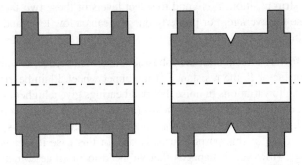

Figure 2.5   Factor C. The profile of the upper wheel can be V shaped or U shaped.

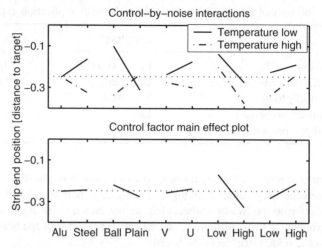

*Figure 2.6   Factor–effect plots for the main effects and control-by-noise interactions.
The centre line is the overall average.*

something about designing a strip applicator that is robust against temperature and
aims at the target value *m*.

(i) First look for control factors that interact with noise factors without having
an effect on the average. Such factors have nonparallel lines in the interaction
plot and flat lines in the main effect plot. From a visual examination of the
graph, there seems to be two such factors in our experiment, namely the strip
tension and the profile of the upper wheel (Figure 2.6). In order to achieve
robustness, select the strip tension at a low level (aluminium weight) and the
profile of the upper wheel at a low value (V shaped).

(ii) Then look for factors affecting the average strip position but not the robustness.
Figure 2.6 reveals that both properties of the lower wheel seem to be such
factors. Since the average in the experiment is smaller than the target value *m*
of the strip position, it is good to select levels of these two factors that will
increase the average. For property one it means a low level and for property
two a high value.

(iii) Some factors affects both the robustness and the average. It seems to be the
case for factor *B*, the resistance of the upper wheel. Plain bearings are more
forgiving to variations in noise, but ball bearings give a higher average. Since
the target value is *m*, we want the average to increase. Thus, a compromise is
needed. Since it is less likely that factors affect the robustness, the few that
can be used for this purpose usually are. In this case it means using plain
bearings. However, it happens that we need to sacrifice something when it
comes to variations in order to aim at the target.

(iv) There might be factors that are included in the experiment that affect neither the robustness nor the average. Such factors are good. They can be used for other purposes, for example reducing the cost. It will soon be seen that one of the factors, namely the upper wheel profile, actually falls in to this category, even if a rough graphical analysis would suggest something else.

Traps in engineering experience

When a DOE for robust design is planned, a major issue is to identify the factors to include in the experiments and their levels. When we select these factors we rely – in some way or another – on engineering experience. However, we all have a tendency to have a much better feeling for factors affecting the average than for those affecting the variation. If there are factors like the strip tension and the profile of the upper wheel that affect the variation but not the average, the risk that they will not be included in the experiment is considerable. The consequence is a missed opportunity for robustness. It also underlines the rule of thumb that factors that can be used to reduce variation should be used for this purpose – even if it means that the average drifts away from the target value.

2.2.3   Analytical analysis and statistical significance

So far, we have only looked at the results graphically; we have not yet asked ourselves if the effects are statistically significant. A simple way to test this is a normal probability plot, as in Figure 2.7. As usual with normal probability plots of factor effects, there is a considerable amount of subjectivity affecting the interpretation. In this case, there are several possible conclusions. The two extremes among them are that none is significant and that all eight marked ones are significant.

An alternative, or complement, to the probability plot is stepwise regression. We will use forward selection with an F test for entering new factors into the model, and a significance level of $\alpha = 5\%$. The sum of squares for the empty model is

$$SS(0) = \sum \left(y_i - \bar{y}\right)^2 = 2.34,$$

where $\bar{y}$ is the overall average. The factor that has the largest correlation with the response y and thus is the first one to enter is the three factor interaction ABN, where N denotes the noise. We obtain the sum of squares

$$SS(1) = \sum \left(y_i - \left(\hat{\beta}_0 + \hat{\beta}_1 x_{ABN}\right)\right)^2 = 2.00.$$

If the contribution from factor ABN is due to random variation, then the ratio

$$F = \frac{(SS(0) - SS(1))/1}{SS(1)/(32 - 2)} = 5.1$$

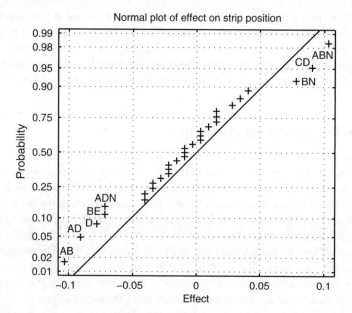

Figure 2.7   A normal probability plot reveals that some of the factor effects may be statistically significant.

is F distributed with $(1, 30)$ degrees of freedom. Since $F = 5.1$ gives the p value

$$p = 0.031,$$

which is less than the significance level α, factor ABN shall be added to the model. We continue and obtain

Step	Factor	p value
1	ABN	$0.031 < \alpha$
2	AB	$0.021 < \alpha$
3	CD	$0.029 < \alpha$
4	AD	$0.019 < \alpha$
5	D	$0.028 < \alpha$
6	BN	$0.017 < \alpha$
7	BE	$0.015 < \alpha$
8	ADN	$0.006 < \alpha$
9	CE	$0.086 > \alpha$

so that the model will include eight terms. These are the eight factors marked in the probability plot (Figure 2.7). It might be a too large a model, partly since F tests are used for entering additional terms. It is well known that the model size has a tendency to be overestimated if F tests are used (see, for example, Hjorth, 1994).

We obtain the model

$$y = 3.25 - 0.08x_D - 0.10x_{AB} - 0.09x_{AD} - 0.07x_{BE} + 0.08x_{BN}$$
$$+ 0.09x_{CD} + 0.10x_{ABN} - 0.07x_{ADN} + \varepsilon,$$

where $\hat{\sigma}_\varepsilon = 0.02$. Three of these terms, x_{BN}, x_{ABN}, and x_{ADN}, are control-(by-control-)by-noise interactions and will be further explored to understand how to obtain robustness.

We have already looked into one of them, x_{BN}. Figure 2.6 reveals that plain bearings make the strip application more robust against temperature. Figure 2.8 helps us to decide the values of the other control factors. The tension-by-resistance interaction, AB, should take the value +1. Since the bearing already is chosen to be on level 1, so must the tension. In practical terms, this means a steel weight. Finally, the interaction between factors C and D should be -1, and thus either a U shaped profile and factor D on a low level or a V shaped profile and factor D on a high level.

Before moving on, a number of reflections on these conclusions are necessary. The first concerns the fact that the experimental design is half fractional. Third order interactions are confounded with other third order interactions. Thus, the interaction ABN is confounded with CDE and ADN is confounded with BCE. As a consequence, we cannot know if the interactions ABN and ADN can be used to accomplish robustness. Thus, the only factor that we know can be used to get robustness against noise is factor B, the rolling resistance.

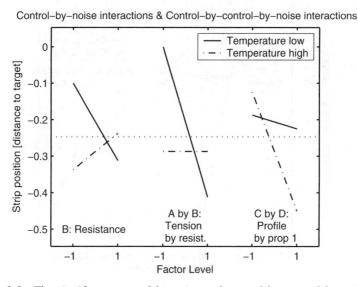

Figure 2.8   The significant control-by-noise and control-by-control-by-noise interactions (BN, ABN, and ADN).

Another reflection is that the presence of a statistically significant control-by-noise interaction and that the control factor involved can be used to accomplish robustness are not two identical things. Consider, for example, the bearing-by-temperature interaction. From the interaction plot it seems that neither ball bearings nor plain bearings make the strip position completely independent of temperature. For plain bearings the position Y seems to increase with higher temperatures, but for ball bearings it decreases. Thus, there is an interaction, but none of the choices is really good (even though the plain bearing seems to be better). Since the factor is not continuous but categorical and thus there is no numerical value of the bearing that it takes where the lines cross each other, the bearing is not necessarily a good factor to get robustness, even if the interaction is significant.

The conclusion is that a combination of analytical and graphical techniques is needed.

2.2.4   Some additional comments on the plotting

In Figure 2.6 the control factors are found on the x axis and the noise factor level is marked by different line types. However, there is another way to plot the same interaction, namely to exchange the place of the noise and the control and have the noise on the x axis (Figure 2.9). Then, we will look for control factor levels giving flat lines. Figure 2.9 and the upper diagram in Figure 2.6 provide us with exactly the same information. It is simply the way it is presented that differs. Which one to choose is a matter of taste.

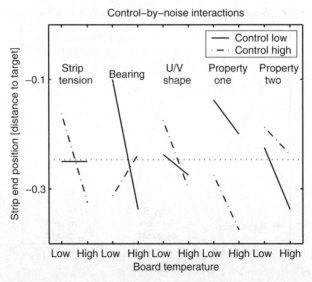

Figure 2.9   An alternative way to plot the control-by-noise interactions.

2.3   Dispersion effects

There are certainly other sources of variation for the strip position apart from the temperature; otherwise it would be unnecessary to have replicates since they would give identical results. Unless these sources are included as factors in the experiment, or at least observed and recorded, we cannot know how they vary during the experiment. Some of the sources might be more or less constant, while some may vary substantially. Just as for the modelled noise, the temperature, this variation will propagate to the response, and once again it may be control factor levels that make the design more robust to this unmodelled noise.

If there are replicates in the experiment, it is possible to explore this. The effect factors have on variability stemming from noise that is experimentally uncontrollable and unobservable is called the dispersion effect. In order to explore dispersion effects, we need to build a statistical model with some kind of measure of dispersion as the response. The most common ones among such dispersion measures are

- the standard deviation σ;

- the variance σ^2;

- the logarithm of the standard deviation $\log(\sigma)$;

- the coefficient of variation σ/μ.

There are mathematical and statistical reasons that make $\log(\sigma)$ a convenient choice in many situations. For example, it is more likely to have an additive model if $\log(\sigma)$ is used. Thus, it will be used here.

For the strip application, there were in fact 10 replicates. The sample standard deviation for each factor combination and its logarithm for this data are given in Table B.3. Just as earlier on, a good starting point is to look at the factor–effect plots (Figure 2.10). The strip tension and the rolling resistance seem to be the ones that affect the variation the most. Once again it is a low value of the strip tension and a high value of the rolling resistance that seem to be the best choice for decreasing the variation.

A normal probability plot can be used to check if the effects are statistically significant (Figure 2.11). There are only two factors that have any significant effect on the dispersion, namely the strip tension and the noise factor, the board temperature. The conclusions on how to design the strip applicator are summarized in Table 2.3.

The decisions taken by the engineers who developed this strip application once their experiment was performed and their data analysed was to go for a small strip tension, that is a strip accumulator of aluminium. As we have seen, the experimental results pointed to plain bearings in order to obtain a strip applicator that can accommodate variations in temperature. However, there were some doubts among the engineers; for example the type of plain bearings used in the experiment may have been relevant, but the not the ball bearings. Thus, the bearings used in the test may

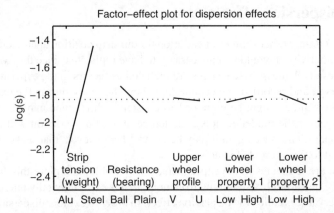

Figure 2.10   A factor effect plot for $\log(s)$ *in the strip position example.*

not be representative. The choice of bearings required more investigation. Finally, the average value of the strip position Y was below the target value m for all tested factor combinations. However, this was not considered as a problem since there is an additional factor, namely a sideway movement of the bracket on which the wheels are mounted. Such a movement has a nearly linear effect on the average strip position Y and was included in some additional experiments and finally used to tune to target.

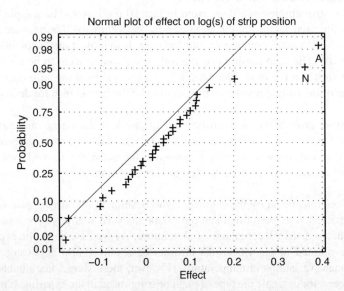

Figure 2.11   The temperature and the strip tension are the only factors that have a significant effect on $\log(s)$.

Table 2.3   The conclusions about how to design the strip applicator, based on the analysis in this chapter, including the result of Exercise 2.1.

Factor	Select level	Which means	Will give
Strip tension	Low	Al weight	Robustness
Rolling resistance	High	Plain bearing	Robustness
(or	Low	Ball bearing	Higher average)
Upper wheel profile	Any	—	—
Property 1 lower wheel	Low	Low	Higher average
Property 2 lower wheel	High	High	Higher average

Exercises

2.1 Consider the strip position problem. In Section 2.3 the fact that there actually are 10 replicates was used, but not in Section 2.2. Use the entire set of data given in Table B.2 to check out which terms are statistically significant. Is it possible to accomplish robustness against temperature?

2.2 For the strip position data of Example 2.1, the temperature is the noise factor. The problem was analysed in order to find out how to design in order to get a strip position that is robust against this noise. The effect of the temperature could first have been estimated in order to judge whether the impact it has on the response is large enough to matter. Estimate the effect of the noise and reflect upon the result!

2.3 Once an experiment is performed, it is a good habit to look at the data graphically to see if there are any data points that look strange. Try to find a way to plot the data of the strip application.

2.4 The change in temperature between a heating element and the next operation in an industrial process depends on a range of factors, such as the distance between the heating element and the downstream operation. There are also a number of noise factors (Table 2.4). The temperature in this downstream operation is important for the result of the operation.
  (a) Analyse the data of Table 2.4. Build a mathematical model.
  (b) Are there any ways to make the temperature change robust against the noise factors?
  (c) In this study, the production rate (units per time unit) was considered as a noise factor. Are there any other alternatives?

2.5 Consider the interaction plot of Figure 2.12. How can the information that the plot provides be used to obtain robustness if:
  (a) The control factor is numerical and continuous, as, for example, a temperature in a production process?

Table 2.4   The temperature drop data of Exercise 2.4.

| Control factors | | Noise factors | | | Response |
Distance A	Factor B	Factor C	Factor D	Production rate (E)	ΔT y
150	30	2	2800	2	6.2
100	20	0	800	1	6.1
200	20	0	800	3	5.6
100	20	4	800	3	9.2
200	20	4	800	1	13.5
100	20	0	4800	3	6.4
200	20	0	4800	1	10.2
100	20	4	4800	1	11.5
200	20	4	4800	3	12.7
150	30	2	2800	2	7.0
100	40	0	800	3	1.6
200	40	0	800	1	4.5
100	40	4	800	1	6.2
200	40	4	800	3	5.9
100	40	0	4800	1	3.8
200	40	0	4800	3	4.3
100	40	4	4800	3	6.3
200	40	4	4800	1	8.2
150	30	2	2800	2	6.4

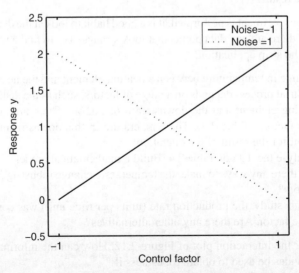

Figure 2.12   A control-by-noise factor interaction plot (Exercise 2.5).

(b) If the control factor is numerical, are any additional experiments valuable before any conclusions are drawn, and how would such an experiment look?

(c) The control factor is nonnumerical, as, for example, a component that can be bought from either one of supplier A or supplier B?

Reference

Hjorth U (1994) *Computer intensive statistical methods. Validation, model selection, and bootstrap.* Chapman & Hall/CRC.

3

Noise and control factors

3.1 Introduction to noise factors

One of the first questions to ask in robust design is what to make robust, the response variable. At first, this question may seem unnecessary and the answer evident. Nevertheless, it is one of the most difficult and crucial questions to answer.

Once the response is defined, it must be made clear what it should be robust against. It is the noise factors, which are the main focus of this chapter.

As Deming said, 'Variation is the enemy of quality'. This is exactly why noise factors are interesting – variation is undesired. However, it is not the variation of the noise itself that is the major concern; it is the propagation of variation from the noise factors to the response. Can this propagation in some way be prevented, as in Figure 3.1?

In Chapter 2, this question of variation propagation was addressed. However, before making anything robust against noise, it must be asked what this noise could be. Understanding this is the key to success.

Noise is a parameter that is not under control during the intended usage of a product, at least not controllable at a reasonable cost. However, the noise must be controllable in the experimentation, otherwise it is impossible to generate knowledge of how to achieve robustness against it (even though some exceptions to this are treated in Chapter 7 – but at the cost of an increased size of the experiment). The response depends on the value taken by the noise factor. In a DOE for robust design, the most important types of noise must be taken into consideration.

Statistical Robust Design: An Industrial Perspective, First Edition. Magnus Arnér.
© 2014 John Wiley & Sons, Ltd. Published 2014 by John Wiley & Sons, Ltd.
Companion website: www.wiley.com/go/robust

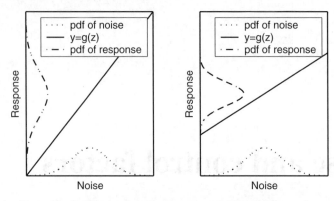

Figure 3.1   The variation in noise will propagate to a variation in the response, but the product can sometimes be designed in such a way that this propagation becomes small, as in the right diagram.

3.1.1   Categories of noise

Noise factors are sometimes classified into different categories in a way that can help to understand the noise better and what it can be. These categories can, for example, be

- external noise;
- internal noise (age);
- manufacturing variation;
- surrogate noise.

This is not the only way to divide noise factors into groups. Especially 'external noise' is very broad and can be further divided into smaller categories.

External noise

There are many different types of external noise. The behaviour of customers and consumers and decisions taken by other groups of people are two examples. A third type, one of the most important, is environmental noise. It can typically include

- variation in the supply voltage to production equipment;
- humidity in a storage area for production material;
- vibrations in a machine;
- electromagnetic disturbances;
- ambient temperature.

Figure 3.2   Decisions taken by someone else can be considered as noise. The position of the rearview mirror may be considered as noise by a noise and vibration engineer.

The noise z is not always a quantity that varies randomly, as an ambient temperature. If it is desired that $y = g(z)$ takes more or less the same value independently of the value of z, then z can be considered as noise. Decisions taken by other groups of engineers can be such a factor. It is certainly nothing that varies randomly, but still something that cannot always be foreseen and hence something to be robust against. There is another reason as well, even more important than the unpredicable behaviour of other engineers. The design freedom will be larger for these engineers if their design choices are considered as noise. Considering their decisions as noise is a way, and a very efficient one, of allowing them to have the widest possible design freedom.

Example 3.1   A team of noise and vibration engineers at an automotive company is interested in the wind noise from the rearview mirror (Figure 3.2). The position (height) and size of the rearview mirror is decided by the ergonomics engineers and the decision may be taken in a late stage when there is no possibility to redesign. Therefore, the team decided to consider this position and size as noise factors and design to become robust against it.

As already mentioned, another type of external noise is customer and consumer behaviour.

Example 3.2   Several lids and knobs, such as the lid to a CD reader, a knob for switching on or off the light, a glove box or a space bar on a typewriter keyboard, are

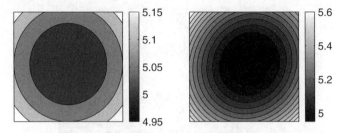

Figure 3.3   The push force required to release the lid as a function of the force application position. The lid to the left is the one more robust against position.

often pushed inwards or downwards in order to be released or activated. However, the user will not push on the same position each and every time. Especially for a space bar on a keyboard, which the writer uses without taking eyes off the text, or for the light button in a dark room, the variation in applied positions can be large. It may be desired that the force needed is robust against this variation in position. The user behaviour is then considered as noise (Figure 3.3).

Example 3.3   The information display in a car can provide the driver with information of the distance remaining until the fuel tank is empty, 'distance to empty'. It is a prediction calculated in the instrument panel electronics and is based on information from the engine control system. This information depends on how the car has been driven so far, and so is the output of the prediction algorithm. The algorithm output is thus highly dependent on driving pattern. However, it is unknown how the car will be driven in the future, so a prediction of, say, 280 km is just a prediction. If it is desired that this prediction is robust against some aspects of the driving pattern of the car, such as if the car is driven on flatland roads first and then in the mountains or vice versa, then this order can be considered as noise. One way to deal with it practically in experiments could be to electronically store the driving pattern of, say, a test car and then divide the time series into pieces and swap the order, either in a structured or in a random way.

Internal noise

An engineer is about to develop a pump. He or she wants the pump to work well, not only at delivery but for a long time thereafter. The properties of the pump should be robust against age. The age of the pump (or of some component of the pump) can then be considered as noise. Once again, this does not vary randomly – the device will get older. The age is sometimes called internal noise.

Manufacturing noise

Manufacturing variation is primarily about part-to-part variation, which is a random variation around the nominal. This noise can to some extent be controlled by the engineer by tightening tolerances, but for an undesired cost. Tightening tolerances may mean more expensive suppliers, investment in new and better production equipment, or slowing down the production rate. This part-to-part variation can typically be

- variation in tensile strength of a plastic component;
- concentration of a certain type of chemical in granules of plastics;
- variation around nominal resistance for mass-produced resistor components;
- variation in thickness from position to position of a metal sheet.

It is worth noticing that the last one in the list, the thickness variation, is not really a part-to-part variation; it is the variation within one and the same part.

Manufacturing noise may also include other things, things that unlike the ones just mentioned can hardly be considered as variation around a nominal value. For example, if two metal sheets are welded together with spot welds, some of these spot welds will be of low quality and some will be totally missing. This is typically the case when welding metal sheets in automotive engineering. When the experimental design is set up, this could be a noise factor where one of the levels is that a certain proportion of the spot welds, say 1%, are missing.

Surrogate noise

In Example 3.1, the wind noise, it was emphasized that noise is not necessarily something that varies randomly, as an ambient temperature. The response should take roughly the same value no matter what value this noise factor takes. Surrogate noise is typically something of this type. Let us study some examples.

Example 3.4   Brake torque. When a torque is applied, this torque will not be a step function, going directly from zero to the target torque. It may be a smooth build-up of the torque or it might look as in Figure 3.4, with a substantial overcast followed by a damped sine curve. The response is a function of time. However, as small a variation over time as possible is desired. Time is in no way random, but it can still be considered as a noise variable.

As noise levels, there are several possible choices, such as

- the largest and smallest values along the curve (A_1 and A_2 in Figure 3.4);
- values at preselected times (T_1 to T_3 in Figure 3.4);
- values at random times.

The best choice is probably the last one of these alternatives since the second may allow a sine wave with the same frequency as the sampling but any amplitude, and the first has other drawbacks.

Example 3.5   A classical example of surrogate noise taken from one of the early robust design applications of Taguchi concerns the baking of tiles in a kiln. There was a variation in the final tile dimensions depending on their location in the stack (Figure 3.5). Tiles at the edges of stack did not conform to the specification. Tiles in the centre of the stack were as specified. The position was considered as a noise factor.

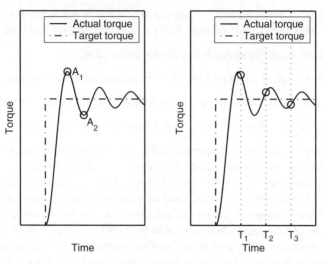

Figure 3.4   Brake torque.

Another type of surrogate noise is the dependence on direction. Such noise is sometimes present in measurement systems. Both hysteresis and friction in the measurement system might cause the read-out value from the measurement system to depend on the preceding measurement. If a certain measurement is preceded by the measurement of a smaller value, a small value is typically obtained and vice versa. The noise can be the direction in which the measurement system approaches the new measurement (Figure 3.6).

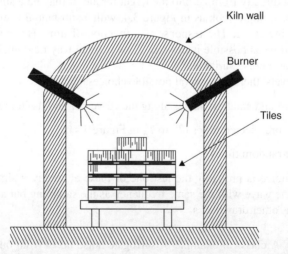

Figure 3.5   The position in the stack while baking tiles in a kiln affects the final dimensions and can be considered as noise. Phadke, Madhav S., Quality Engineering Using Robust Design, *1st Edition, © 1989. Reprinted by permission of Pearson Education, Inc., Upper Saddle River, NJ.*

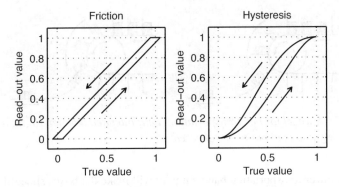

Figure 3.6   The direction in a measurement system can be considered as noise. The read-out value will depend on the preceding measurement.

We have now seen examples of three types of surrogate noise, namely

- time-to-time;

- position-to-position;

- direction-to-direction.

The word 'surrogate' suggests that this kind of noise is not a kind of noise in itself but only a way to simulate variation 'without necessarily controlling the source of variation', as Fowlkes and Creveling (1995) states. Take, for example, the tile baking example. The noise may be a variation in temperature (the heat does not fully reach the centre of the stack) or exposure time (there is a delay before the heat reaches the centre so the exposure time is shorter), or even in the time to cool down. The temperature (or time) may then be seen as the true noise and the position only as a way to simulate it. By having this view on surrogate noise, replicates can be included in this category as well.

3.2   Finding the important noise factors

3.2.1   Relating noise to failure modes

In the Taguchi School of robust design, the response variable should reflect the intention of the design, so minimizing a failure mode is not a primary goal in the Taguchi type approach. There are several reasons for this, as will be discussed in Chapter 6. However, even if the failure mode is not used as a response, it can be valuable to pay attention to it in some manner. It could be through the noise factor, as Example 3.6 shows.

Example 3.6   Consider a wheel that transports boxes as in Figure 3.7. An upright box comes to the wheel that has a number of racks or shelves and moves the box

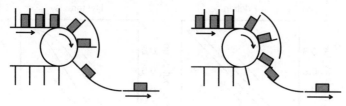

Figure 3.7   A wheel for box transportation. Left: intended manner. Right: a box slips off too slowly and is about to be squeezed.

down to a lower transportation band. Gravity will make the boxes slide off the rack. There is a protecting shield preventing the boxes from flying off the wheel too early due to centrifugal forces.

If the box does not slip off fast enough, it may be squeezed by the next rack. The slip-off time depends on the friction between the shelf or chute and the box. If the shelf, chute, and box are wet, this will affect the friction. Therefore, dry or wet can be included as a noise factor. In that way, the failure mode, the squeezed box, is considered by including its cause as noise in the experiment.

Even if no jammed boxes are observed in the test (where the sample size is limited), it is more likely to be some jammed boxes on the field if the variation in the slip-off time is high.

3.2.2   Reducing the number of noise factors

If the number of noise factors is large, the number of runs in the experiment easily become too large to handle. In this subsection, two methods are studied for reducing the number of noise factors to include in the experiment by picking out the few vital ones, namely

- screening experiments;
- variation mode and effects analysis (VMEA).

The two approaches are very different from each other. The name of the first one, the screening experiment, points out that it is an experiment in itself, one that precedes the main experiment. In contrast to this, the second one (VMEA) is an analytical and qualitative method. They both have drawbacks and advantages and do not necessarily exclude each other. The VMEA could be a first step and the screening experiment a second step in the identification of noise factors.

Screening of noise factors

The process of learning is stepwise and interactive; we run a test and learn something that we want to explore further. Another test is run and new conclusions are drawn, some just confirming former conclusions, some giving birth to new ideas, and some causing adjustments and refinements of already available knowledge. In that way, the natural approach in all experimentation is sequential testing. Robust design is in no way an exception to this and screening experiments of noise factors is a nice example of it.

As already mentioned, the problem that is addressed in a noise factor screening experiment is that the list of potential noise factors is so long that all of them cannot be included in the main experiment. The approach is the same as for screening experiments in traditional DOE so the readers who are familiar with this will not find the next few pages surprising. However, there is one issue that makes it more complicated. The issue is that the key to robustness is in the interactions, but a screening experiment has a tendency to focus on the main effects.

The steps of a noise factor screening experiment can be:

(i) Select one control factor combination.

(ii) Set up a highly fractionated array for noise factors only with as few levels for each factor as possible.

(iii) Perform the experiment.

(iv) Analyse the data and pick out the few vital ones.

The type of experimental array that is selected depends on the original number of noise factors. If there are seven of them, a 2^{7-4} array may be a good choice, and if there are 11 original noise factors, a so-called Plackett–Burman design is one to consider. They have in common the fact that only main effects can be estimated, and only if it is assumed that all the interaction effects are of negligible size.

The first step in the noise factor screening experiment, the one where the control factors are fixed, has one important drawback that can jeopardize the entire robust design study. The idea of robust design is to exploit control-by-noise interactions. A function can be made robust if the impact of a certain noise factor is large for some selections of control factor levels and small for other selections of levels. This will complicate the screening experiment and may cause some important noise factors to be left out of the main experiment. To illustrate this, consider a situation with three candidate factors, one control factor A and two noise factors B and C. If the control factor A in a screening experiment is fixed to a value making the response y insensitive to noise factor B, then the conclusion of the screening will be to exclude factor B from the main experiment. In the main experiment, the control factor A will then be used in order to obtain robustness against the noise C. However, this may mean that factor A is set to a value making the design highly sensitive to the noise B, a noise factor that was discarded as unimportant earlier in the screening.

There is no obvious remedy for this. Using experimental arrays of resolution V in a screening experiment is hardly ever an option, as it would then not be small enough for a screening. To include control factors in the screening and thus have a common screening experiment for both types of factors will help a little, but since the focus is still on the main effects it will not remove the basic problem. Nevertheless, common screening of noise and control factors is a good idea. Somewhat reluctantly it may have to be accepted that this drawback exists and that there is considerable room for mistakes. It is not every robust design study that will succeed. The only alternative (at least theoretically) would be to run one really large main experiment. Therefore it all goes back to a choice between one large and expensive experiment, which may not even be run due to the costs involved, and a series of two smaller ones, a screening and a main experiment, with some room for mistakes.

Variation modes and effects analysis

Variation modes and effects analysis (VMEA) is a top-down method to robust design and includes not just the identification of noise factors but also guidance for where and for which functions robust design should be applied. It is covered in more detail in Chakunashvili *et al.* (2009) and Johansson *et al.* (2006), the original articles of VMEA. Like so many other top-down methods it has an unfortunate tendency to move the activities from actual engineering to endless meetings in conference rooms. Therefore, I am personally somewhat reluctant to use it. Despite this drawback it certainly has its benefits and its rightful place in this chapter.

From a mathematical point of view, the VMEA is just like so many other aspects in robust design nothing but an application of a second order Taylor expansion (Section 1.3). Extending the view beyond the mathematics, it is also something more, namely a structured way to prioritize noise factors as well as a way to point out where, for which so-called sub-key product characteristic (KPC), to apply robust design.

The response, the key product characteristic (or key process characteristic depending on the application, but still with the same abbreviation), is affected by a number of sub-KPCs, which are in turn affected by noise. This can be represented in terms of a fishbone diagram (Figure 3.8).

The VMEA ends in a prioritization. It will point out which sub-KPCs are the most important ones when it comes to variation of the KPC and it will point out noise factors that are the most important ones both with respect to the sub-KPCs and to the KPC. The first part, the prioritization of sub-KPCs, gives guidance on the selection

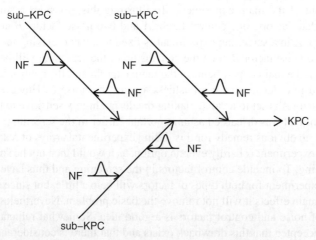

Figure 3.8   A fishbone diagram for VMEA.

of where to apply robust design. The second part, the noise prioritization, indicates which noise factors to include. Here, the VMEA is structured into a set of 10 steps.

Step 1. Define a set of product characteristics.

Step 2. Select the one (or the ones) that are of particular interest from the variation point of view (KPCs).

Step 3. The KPC is cascaded into a number of sub-KPCs whose values affect the KPCs. A sub-KPC is an engineering attribute, just like the KPC. If the VMEA is applied to a bolt and nut fastener and the KPC is the difference in diameters between the bolt and the nut, a sub-KPC could be the outer diameter of the bolt and another one the inner diameter of the nut. If, in another problem, the KPC is the grip stiffness of a milk package, the sub-KPCs might be surface friction, fill volume, and board thickness. The property that all these sub-KPCs have in common is that they are measurable quantities. It is an important property for a response variable in robust design, as will be discussed in Chapter 5.

Step 4. Each sub-KPC is associated with the potential noise factors affecting it. After step 4, a stage is reached where it is possible to sketch a fishbone diagram as in Figure 3.8. The breakdown of the KPC into sub-KPCs and further into noise factors really facilitates the understanding of variation.

Step 5. In a qualitative judgement, the influence U_S of each sub-KPC on the KPC is assessed together with the impact U_N of noise factors on each sub-KPC. A set of criteria for these judgements (Table 3.1) has been developed by Johansson *et al.* (2006).

Table 3.1 Ranking scale for the influence that a noise factor is judged to have on the sub-KPC (or the sub-KPC on the KPC). Scales taken from Johansson et al. (2006).

Sensitivity	Criteria for assessing sensitivity	Score
Very low	The variation of noise (alternatively of sub-KPC) is (almost) not transmitted at all to sub-KPC (alternatively to KPC)	1–2
Low	The variation of noise (alternatively of sub-KPC) is transmitted to sub-KPC (alternatively to KPC) to a small degree	3–4
Moderate	The variation of noise (alternatively of sub-KPC) is transmitted to sub-KPC (alternatively to KPC) to a moderate degree	5–6
High	The variation of noise (alternatively of sub-KPC) is transmitted to sub-KPC (alternatively to KPC) to a high degree	7–8
Very high	The variation of noise (alternatively of sub-KPC) is transmitted to sub-KPC (alternatively to KPC) to a very high degree	9–10

Table 3.2   Ranking scale for the (guessed or expected) variation of the noise factors. Scales taken from Johansson et al. (2006).

Variation of noise	Criteria for assessing the variation of noise	Score
Very low	Noise factor is considered to be almost constant in all possible conditions	1–2
Low	Noise exhibits small fluctuations or lies within a small interval in all possible conditions	3–4
Moderate	Noise exhibits visible but moderate fluctuations in all possible conditions	5–6
High	Noise exhibits visible and high fluctuations in all possible conditions	7–8
Very high	Noise exhibits very high fluctuations in all possible conditions	9–10

Step 6. Assess the magnitude of variation V of noise factors in operating conditions. A possible ranking scale is given in Table 3.2. The ranking criteria are summarized in Figure 3.9.

Step 7. Calculate the 'variance' for a sub-KPC. If it is affected only by one noise factor it becomes

$$U_N^2 \times V^2$$

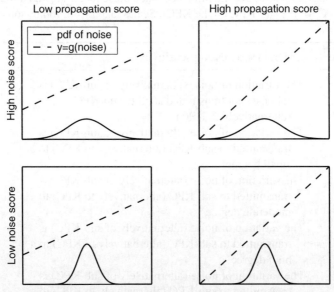

Figure 3.9   A summary of the score in VMEA.

and if it is affected by several noise factors

$$\sum_{\text{all noise}} U_N^2 \times V^2.$$

It should be obvious from the discussion in Section 1.3 why the squares of the terms are used.

Step 8. Calculate the variation risk priority number (VRPN) for the effect that a sub-KPC has on the final KPC:

$$VRPN_{\text{sub-KPC}} = U_S^2 \times \sum \left(U_N^2 \times V^2 \right).$$

Step 9. Calculate the variation risk priority number for the effect a noise has on the final KPC:

$$VRPN_{\text{noise}} = U_S^2 \times U_N^2 \times V^2.$$

Step 10. Use $VRPN_{\text{sub-KPC}}$ to choose what to apply robust design on (if not the KPC itself) and $VRPN_{\text{noise}}$ to select the noise factors that will be included in the robust design experiment.

A VMEA is mostly based on judgements rather than facts. Even if it is the judgements of skilled and experienced engineers, they are still just judgements. In the worst case it may even be speculation and speculation is something that we have too much of in industry. Therefore, the VMEA is just a tool to identify the noise factors that are supposed to be the most important ones and will be included in the experiment, as stated in step 10. It is not the end product.

The mathematics of VMEA

If the relation between a noise factor z_{ij} and a sub-KPC y_i is

$$y_i = g(z_{ij})$$

the propagation from the noise to the sub-KPC will consist of two elements

$$\left(\frac{\partial g_i}{\partial z_{ij}} \right)^2 \sigma_{z_{ij}}^2$$

in accordance with the discussion in Chapter 1. If there are several noise factors affecting one and the same sub-KPC, they sum to

$$\sigma_{y_i}^2 = \sum \left(\frac{\partial g_i}{\partial z_{ij}} \right)^2 \sigma_{z_{ij}}^2.$$

This variation will then propagate one step further, to the KPC itself,

$$KPC = h(y_1, \ldots, y_m),$$

according to

$$\sigma^2_{KPC} = \sum_{i=1}^{n} \left(\left(\frac{\partial h}{\partial g_i} \right)^2 \sum_{j=1}^{m_i} \left(\left(\frac{\partial g_i}{\partial z_{ij}} \right)^2 \sigma^2_{z_{ij}} \right) \right),$$

where n is the number of sub-KPCs and m_i the number of noise factors affecting the ith sub-KPC. Since it is mostly the case that neither the noise variation $\sigma^2_{z_{ij}}$ nor the sensitivity factors $\partial g_i / \partial z_{ij}$ and $\partial h / \partial g_i$, are known, this quantitative approach to VMEA usually exchanged for a qualitative one, the one with the scales of Tables 3.1 and 3.2.

3.3 How to include noise in a designed experiment

3.3.1 Compounding of noise factors

Suppose that a cap to a bottle is about to be designed with the aim that it should be easy to open. Let us say it is a cap with a so-called tampering band (Figure 3.10) and that the quantity to make robust is the peak force while opening, which is supposed to be closely related to what customers mean by easiness to open. The noise factors are the variation around the nominal bridge width (the small bridges between the tampering band and the main part of the cap) and the shrinkage of the cap. All moulded plastic parts will shrink and their final dimensions will thus be smaller than those of the mould cavity. The amount of shrinkage is typically a common response in robust design for a moulding process, and a typical noise factor for functions coming downstream of the moulding, in later manufacturing steps and in product usage.

Figure 3.10   The bridge width is not the same for every cap. A wider bridge is likely to cause higher opening forces.

A really interesting property of the noise may be known even before the start of the experiment, namely its 'directionality'. Directionality is certainly not a word used in everyday English, but the expression 'directionality of noise' has a very specific meaning in robust design; it is known that a wider bridge will increase the opening force, and so will a higher value of the cap shrinkage (it will be tighter between the cap and the neck). Thus, the directionality is known. If this directionality is known, we can compound (group) the noise factors.

Definition 3.1 (Directionality) *Let*

$$y = g(x_1, \ldots, x_p, z_1, \ldots, z_q) \tag{3.1}$$

be the quantity that should be robust, x_i, $i = 1, \ldots, p$, control variables, and z_j, $j = 1, \ldots, q$ noise variables. Then if

$$u_k = \frac{\partial g}{\partial z_k} > 0 \tag{3.2}$$

for all values of x_i and z_j then the directionality of z_k on g is positive. If

$$u_k = \frac{\partial g}{\partial z_k} < 0 \tag{3.3}$$

for all values of x_i and z_j then the directionality of z_k on g is negative.

If the directionality of more than one noise factor is known, noise factors can be compounded (also called grouped). If noise factors are compounded, only the combinations of noise giving the most extreme values of the response are considered. For the opening force, a small shrinkage and a thin bridge would be one level of the compounded noise, and large shrinkage and wide bridge the other level. The other two combinations are left unconsidered.

The goal is to bring the opening force for the two extremes closer together. If this is accomplished, it is likely that the points located in between the two extremes will come closer as well. With the cap opening and the compounded noise of Table 3.3, the points to consider in experimentation are marked in Figure 3.11.

The reason for compounding is to bring down the size of the experiment. However, there is a drawback. By compounding, there is some knowledge that will be impossible

Table 3.3   Table of compounded noise.

Compounded noise at level −1	Compounded noise at level 1
Small shrinkage	Large shrinkage
Narrow bridge	Wide bridge

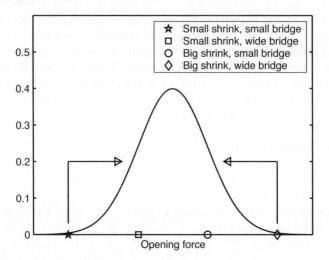

Figure 3.11   The opening force has a tendency to be small if the bridge width is small and if the shrinkage is small, and large if these noise factors are large. By compounding the noise factors, only the extremes are considered.

to obtain. There will not be any opportunity to estimate the effects of the individual noise factors. Knowledge about interactions will also be lost. If control factor *A* interacts with the compounded noise and can be used for making the design more robust, only that is known. We do not know if factor *A* makes the force insensitive to the bridge width, the shrinkage, or both. Nevertheless, the design can be made equally robust no matter if the noise is compounded or not – if the compounding is done correctly – just that it is not known exactly why.

Thus, compounding is beneficial for experimental size but negative for engineering insight.

There are situations when compounding is not an option. One is when the directionality of the noise is unknown. A way to overcome this issue is to perform an experiment prior to the robust design experiment to investigate this. It would be a small designed experiment that is used to sort out the directionality. Even if it adds an additional experiment, the total amount of test runs may be smaller.

Another situation that can remove the possibility to compound noise factors is if the relation between noise and response is not monotone (Figure 3.12). However, even if these factors are not compounded with anything else, they are difficult to handle in the experiments since they mostly require more than two levels in the tests.

Example 3.7   Rubber bushings are used in automobiles in order to dampen vibrations. A high dampening is good from a comfort point of view as well as for audible noise, but from a handling point of view a small dampening is preferred. Consequently, there is a target value that gives a good compromise. Suppose that there are three noise factors, the bushing age, the diameter, and the bushing hardness

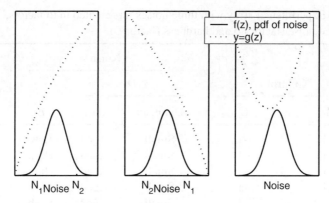

Figure 3.12   The two noise factors to the left can be compounded, but not the one to the right.

(its shore value). The directionalities of the first two of them are known; the dampening property will decrease with age and increase with diameter. Hence, they can be compounded. Under some circumstances it may also be possible to include the hardness in the compounded noise. However, if there are uncertainties, the hardness must be treated separately. Suppose that a 2^{5-1} factorial design with three control factors is set up to explore the dampening property and that a geometry factor (factor A), bushing thickness (factor B), and rubber recipe (factor C) are the control factors (Table 3.4, artificial example). Look specifically at how the compounded noise (factor D) is handled in the experimental array.

Selection of noise factor levels

Montgomery, one of the forefront figures in engineering statistics, once wrote that two-level factorial designs should be the cornerstone of DOE in the industry (Montgomery, 2005). I agree (with the exception of virtual experiments), and this also applies to DOE for robust design. However, even if two-level designs are the backbone, there are occasions when some other experimental designs come better to hand. For example, if substantial nonlinearity is expected, several levels might need to be used. Even if the number of levels is limited to two, the values of the levels are not easily selected.

Consider Figure 3.13. Assume that the random variation of the noise in real life, the density curve of the figure, is known. Then, if the levels are selected too close to each other, the real world variation will not be reflected and the data analysis may point out the impact of the noise as insignificant. If the levels are selected too distant from each other, the problem may be the opposite, that the impact of this noise factor overshadows everything else. Selecting the levels too far apart is related to another risk, namely that nonlinearities outside the region of interest can affect the conclusions.

To choose the values of the levels, it is not required that the exact probability distribution of the noise factors is known. The levels N_1 and N_2 do not necessarily

Table 3.4   Experimental array with three noises represented in to factors,
compounded noise (factor D) and hardness (factor E).

				Noise			
	Control				D		Response
Run	A	B	C	Diameter	Age	E	y
1	−1	−1	−1	Small	Aged	Low	10.2
2	1	−1	−1	Small	Aged	High	10.2
3	−1	1	−1	Small	Aged	High	10.4
4	1	1	−1	Small	Aged	Low	10.7
5	−1	−1	1	Small	Aged	High	8.5
6	1	−1	1	Small	Aged	Low	8.9
7	−1	1	1	Small	Aged	Low	9.5
8	1	1	1	Small	Aged	High	9.1
9	−1	−1	−1	Large	New	High	9.0
10	1	−1	−1	Large	New	Low	9.1
11	−1	1	−1	Large	New	Low	8.9
12	1	1	−1	Large	New	High	8.9
13	−1	−1	1	Large	New	Low	11.5
14	1	−1	1	Large	New	High	11.3
15	−1	1	1	Large	New	High	11.9
16	1	1	1	Large	New	Low	12.0

need to be selected symmetrically (Figure 3.14). However, there must be a rough idea
of the variation range of the noise. If, say, the temperature in a factory is a noise factor
in a certain experiment, then it might be sufficient to know that it is mostly between
10°C and 50°C. However, if 10°C and 50°C is the true but unknown range but the
experimenter believes that this temperature is in the span 20°C to 22°C or in the span
−20°C to −10°C, then this knowledge is insufficient to select levels in a good way.

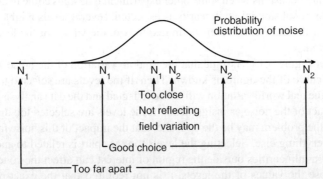

Figure 3.13   The levels of noise should not be too close to each other and not too
far apart.

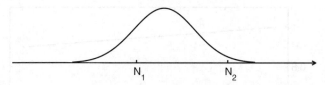

Figure 3.14   It is usually not a problem if the levels of the noise are selected somewhat unsymmetrically.

There is a reason for the probability distribution of the noise not needing to be fully known. In robust design we just try to reduce the variation as much as possible in order to find the design that is the one most likely to fulfil the requirements. The true values of the variation are not under study.

3.3.2   How to include noise in experimentation

Even if noise cannot be affected in real life, it must be possible to affect it in the experimentation. It is only possible to make the design robust against noise factors that are changed and take different values in the experimentation. How can this change of factor levels be accomplished? Let us look at some examples.

Noise in experimentation – some examples

Example 3.8   Slip-off time. In Figure 3.7 we had an example with boxes slipping off a wheel. In this case, the boxes should be on equal distances from each other on the transportation band. The engineers realized that the distance between the packages is related to the slip-off time from the shelf on the wheel and decided to make use of this time as the response. The slip-off time, in turn, depends on friction between the box and the shelf. There are many ways in which the friction can be changed; surface treatment and lubrication are two examples of many. However, the easiest way to change the friction is probably to run it dry or wet, that is to use water as lubrication. It might be sufficient to dry the shelves with a towel and hair-dryer to get them into the dry state and pour some water on to them to get them into the wet state. The message is that even if the change of a noise factor, as the friction, can be accomplished in several ways, it is mostly sufficient to choose the simplest one.

However, it is not necessarily the case that this is good enough. The repeatability of the test might be too poor. The shelves may not be equally dry each time a run with dry shelves is carried out. Something more elaborate may be needed.

Example 3.9   Age. Assume that the age of a nozzle is a noise factor.

This means that new nozzles as well as aged nozzles are needed to perform the test. It is not always an option to go to the customers of the nozzles, for example the plants, and ask for used nozzles. The nozzles collected in such a way would be used in different environments, under different loads and to different degrees so

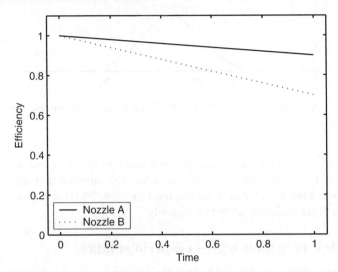

Figure 3.15   A property of nozzles A *and* B *is tested. The test is done for new and for aged nozzles. In the diagram, it is obvious that nozzle* A *is less sensitive to age than nozzle* B, *and is thus a more robust design.*

the tests would not be repeatable. In addition, the approach is not useful at all for new development. Some other type of approach is required.

An accelerated life test can be a good way. The nozzles that will be used in the tests with the noise factor 'age' on a high level are run through an accelerated life test before the robust design experiment. The accelerated life test does not necessarily have to be a perfect acceleration of reality. The intention is not to run to failure. The intention is rather to run during a prespecified time so that the nozzles become somewhat degraded. Nozzles under some settings of the control factors may degrade more than others, and the robust design test will (hopefully) provide such information.

For example suppose that some property of a nozzle shall be robust against age. There are two types of nozzle to choose between, type *A* and type *B*. For each type, this property is measured for some new units and some that have gone through an accelerated life test are measured. The outcome of such a test may look like the one in Figure 3.15 and Table 3.5.

The problem in this example, of how to include age as a noise factor, is to develop a relevant accelerated degradation test.

Example 3.10   Age. Consider a chain or a belt that is used to drive some wheel (Figure 3.16). When the chain becomes older it typically gets longer and longer. If the age is considered as a noise factor, it may be sufficient to test with chains of two different lengths, the nominal one and one slightly longer, and consider this as the two levels of the noise. Even if the chains have not been aged at all, it might be sufficient to consider the effect of age, namely the chain length.

Table 3.5   The results of the tests of nozzle efficiency using age as a noise factor.

Nozzle type (control factor)	Gone through accelerated degradation test (noise factor)	Efficiency (response)
A	No	1.02
B	No	0.97
A	Yes	0.91
B	Yes	0.67
A	No	0.99
B	No	1.00
A	Yes	0.90
B	Yes	0.74

Example 3.11   Injection moulding is a very common manufacturing method. Most of the plastic parts that we come across in our daily life are moulded, from plastic bottles and toys to instrument panels in cars. In injection moulding plastic melt is pressed into a cavity (with the form of the intended object) through one or several gates (Figure 3.17). Assume that, for a certain type of part, there are several gates into the cavity. Sometimes, one or two of these gates are clogged, so that no melt can come through. There are two fundamentally different ways to consider these clogged gates, and which perspective to take depends on what we want. If we want the moulding to be equally good, no matter if some gates are clogged, then this can be considered as noise. On the other hand, if we want to detect a clogged gate as quickly as possible so that it can be cleaned, then it is not a noise factor and the problem is not a problem of robust design; it is to design an efficient detection system.

Assume that we want the moulding to be equally good regardless of whether some gates are clogged. In a robust design test, one noise level could be that no gate is clogged and the other level that one gate is clogged. In the experiment, we must then find a method to clog the gate on purpose. Other issues to consider include which gate to clog and how to make certain that another gate is not clogged unintentionally. If clogging can happen in real life, then it might happen in the laboratory as well.

Figure 3.16   An electrical motor, a drive belt or drive chain, and a wheel (Example 3.10).

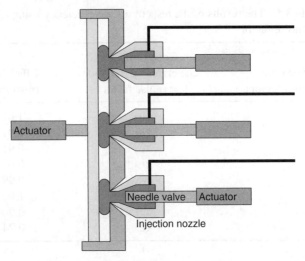

Figure 3.17   In moulding, a plastic melt is pressed into a cavity through one or several gates. A clogged gate might be considered as noise.

3.3.3   Process parameters

A process parameter can be both a control factor and a noise factor, the nominal value the control and the variation around it the noise factor.

Example 3.12   Consider the operation of welding two plastic parts together. The temperature is an important property in the welding process. Suppose that a manufacturing engineer is about to perform a robust design experiment on this with the following factors and levels:

- A: Temperature (control factor), levels 1 and 2.
- B: Pressure applied (control factor), yes or no.
- C: Temperature, variation around nominal (noise factor), levels -0.1 and 0.1.

During the welding operation, the parts may be pressed together or just hold together. The second factor reflects this. The benefit of a pressure is to ensure contact, the drawback being that the melt may be pressed away from the welding point. Even if there are three factors at two levels each, there will only be two settings that are changed in the experiment, namely the temperature and the pressure. The temperature will take four different levels, as given in Table 3.6.

3.4   Control factors

Control factors are factors

- whose values can be decided by the engineer;
- that are expected to affect the response.

Table 3.6   If a factor is both control and noise, the number of parameters that are changed in the experiment is smaller than the number of factors in the experimental array.

| Coded units | | | Uncoded units | | | Machine settings | |
A	B	C	Nominal temperature	Pressure	Temperature variation	Temperature	Pressure
−1	−1	−1	1	No	−0.1	0.9	No
1	−1	−1	2	No	−0.1	1.9	No
−1	1	−1	1	Yes	−0.1	0.9	Yes
1	1	−1	2	Yes	−0.1	1.9	Yes
−1	−1	1	1	No	0.1	1.1	No
1	−1	1	2	No	0.1	2.1	No
−1	1	1	1	Yes	0.1	1.1	Yes
1	1	1	2	Yes	0.1	2.1	Yes

There is nothing strange about such factors. We are used to them from any kind of designed experiments, where they are treated just as factors (in contrast to noise factors, which sometimes are considered as blocking factors in traditional DOE but are mostly set aside).

Typically, control factors can be

- nominal values of geometrical measures;

- nominal values of material parameters;

- type of algorithm in control software;

- parameter values in software;

- choice of supplier for a valve.

Exercises

3.1 What could the noise factors be in these cases?
  (a) You are about to design a screw cap to a bottle. You want it to be easy to open and have chosen to look at the opening energy as the response. What could the noise factors be? (Make some assumptions about what you can affect and what part of the cap design you work with.)
  (b) Two pieces of wood are glued together in a furniture factory.
  (c) The strip position of Example 2.1 in addition to the temperature that has already been considered in the example.
  (d) A finite element simulation method for the maximum stress of a bracket (before making the bracket robust we want the simulation method to be robust, so this is the task).

Table 3.7   Data from the injection moulding with or without clogged gates. Factors
A, B, and C are control factors.

Run number	Control factors			Noise factors		Response
	A	B	C	Gate 1 clogged	Gate 2 clogged	y
1	−1	−1	−1	No	No	11.21
2	1	−1	−1	No	Yes	10.21
3	−1	1	−1	No	Yes	9.89
4	1	1	−1	No	No	10.72
5	−1	−1	1	No	Yes	9.93
6	1	−1	1	No	No	10.67
7	−1	1	1	No	No	11.17
8	1	1	1	No	Yes	10.36
9	−1	−1	−1	Yes	Yes	8.43
10	1	−1	−1	Yes	No	10.29
11	−1	1	−1	Yes	No	9.83
12	1	1	−1	Yes	Yes	9.44
13	−1	−1	1	Yes	No	9.56
14	1	−1	1	Yes	Yes	9.66
15	−1	1	1	Yes	Yes	8.46
16	1	1	1	Yes	No	10.27

(e) Self learning software. Some vehicles have a rain sensor. It is a little device
that can be mounted close to the rearview mirror in the centre of the car.
It sends out a signal and measures how this signal is reflected back to
the sensor. The software interprets a sudden change of the reflection as a
presence of rain-drops on the windshield and the wipers start. Since the
windshield not will look exactly the same when it is aged as when it is new
(there will, for example, be scratches), the software must be self learning so
that it 'knows' what a normal state is. A team of engineers wants to develop
robust software for this, but what should the software be robust against?

(f) Something from your own field of work.

3.2   In Exercise 3.1, noise factors were identified. For one of the problems there:

(a) Use your engineering knowledge and experience to judge whether anything
can be compounded.

(b) Use your knowledge and experience to judge whether two levels are sufficient for all the noises or if more levels need to be used.

(c) If possible, decide upon approximate levels for the noise factors in the
experimentation.

(d) Point out noise factors that have been identified, but where further investigations are necessary before the three first questions can be answered.

3.3 Recall Example 3.8, the slip-off time of the boxes. As mentioned, the noise factor might be difficult to control even in the test environment and the method described there may be too inaccurate. Try to find some way to control this noise factor in the test to make the test repeatable enough.

3.4 Analyse the data of Table 3.5.

3.5 One way to set up the experiment with the injection moulding of Example 3.11 could be the one shown in Table 3.7. Analyse the data.

3.6 In Example 3.9, the robustness to age for some property of a nozzle was discussed. How many units (nozzles) are needed for the test?

References

Chakhunashvili A, Barone S, Johansson P, and Bergman B (2009) Robust product development using variation mode and effect analysis. In *Robust design methodology for reliability. Exploring the effects of variation and uncertainty* (eds Bergman B, de Maré J, Lorén S, and Svensson T), pp. 57–70. John Wiley & Sons. Ltd.

Fowlkes WY and Creveling CM (1995) *Engineering methods for robust product design.* Addison-Wesley.

Johansson P, Chakhunashvili A, Barone S, and Bergman B (2006) Variation mode and effect analysis: a practical tool for quality improvement. *Quality and Reliability Engineering International,* **22**, 865–876.

Montgomery DC (2005) *Design and analysis of experiments,* 6th edn. John Wiley & Sons Inc.

Phadke MS (1989) *Quality engineering using robust design.* Prentice Hall.

4

Response, signal, and P diagrams

Small variation is a key engineering attribute for achieving high quality products, not to say the key engineering attribute. It may be the variation of the strip position as in Example 2.1, of the filling volume of a bottle, the shrinkage of a moulded part, or the force required to press a button on a keyboard. It is this quantity that should be robust against noise and is our response. To be able to learn how robustness can be achieved, the response must be possible to observe and measure in experiments.

Besides the requirement that the response must be measurable, it is good if it reflects customer wants, is based on the intention of the product, and that failure modes are avoided as responses since they are simply symptoms of nonrobustness. Several of these statements have their origin in the ideas of Taguchi. They will be touched upon over and over again in this chapter. However, first we will look into two different situations of what to make robust.

4.1   The idea of signal and response

Suppose that the brake pedal on a car is applied with a certain force. At some later occasion it is applied once again with the same force. Does the car behave the same in terms of deceleration and stopping distance? Probably not. The deceleration is affected by variation. This is a problem because it makes the brakes unpredictable. Robust design is one way to get it as predictable as possible by making it less sensitive to factors that we as engineers cannot affect. It can be environmental conditions such as temperature and humidity, or the driving pattern before the braking event takes place.

Statistical Robust Design: An Industrial Perspective, First Edition. Magnus Arnér.
© 2014 John Wiley & Sons, Ltd. Published 2014 by John Wiley & Sons, Ltd.
Companion website: www.wiley.com/go/robust

The issue with the brakes differs from what we have studied so far since the target is not a single value, as for the strip application of Chapter 2. Rather, it is a relationship between an input variable, the pedal force, and an output, the deceleration or the stopping distance.

The input, in this case the pedal force, is called the signal and the relation between the signal and response is called the ideal function in Taguchi's terminology.

4.1.1   Two situations

Thus, there are two different situations, as illustrated in Figure 4.1. One is when a response value should have as small a variation as possible around a target. All examples so far in this book have been of this type. The position of the strip of Example 1.1, the temperature drop in Exercise 2.4, and the distance between boxes of Example 3.6 are all examples of this. However, the brake pedal illustrates another situation, a situation when the target is not a single value but a relation between a signal M and a response $y(M)$. In Taguchi's terminology, this is a dynamic model. Other examples when this situation occurs may include

- water tap: signal = angle of turning, response = flow rate;
- autoclave: signal = added energy, response = sterilization efficiency;
- mercury thermometer: signal = temperature, response = length of mercury bar.

In its simplest form, this relation is linear,

$$y = \beta M$$

and the smaller the variation is around this regression line the better. In some cases, as with the mercury bar and the sterilization, the slope β should be as high as

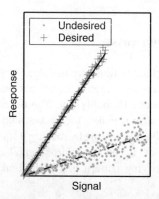
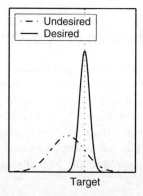

Figure 4.1   In a dynamic system, a high efficiency and a small variation is desired. In a nondynamic system, a small variation and aiming at the target is desired.

possible. For the mercury bar it makes it easier to distinguish two temperatures. For the sterilization, it will reduce the amount of energy that is needed. A design that is both consistent (small variation) and efficient (high slope) is desired.

Most of this chapter is devoted to the dynamic model since it is both harder to understand than a nondynamic model and, according to my personal experience, more likely to give a useful result.

4.2   Ideal functions and P diagrams

A typical example of a dynamic model is a table fan. Mechanical power, rpm times torque, makes the blades rotate to create a flow of air. All the factors, the signal, the noise, and the control, can together with the response be sketched in a P diagram (parameter diagram), as in Figure 4.2.

The signal, noise, and response factors are sometimes called the 'customer space' and the control factors the 'engineering space' since the control factors are considered to be the only ones that the engineer can affect (Figure 4.3). This terminology makes sense for the brake system; it is the customer who applies the force on the pedal. Despite this, the wording 'customer space' is not fully desirable since there are many examples where the signal really is something that the engineer but not the customer can affect.

It is not always easy to understand the difference between a signal factor M and noise and control factors. An explanation can be helpful.

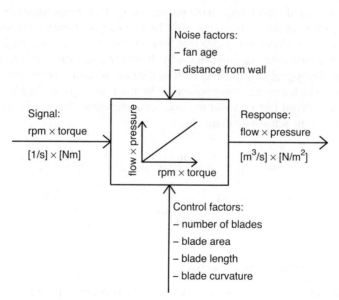

Figure 4.2   A P diagram for the blades on a fan might look like this.

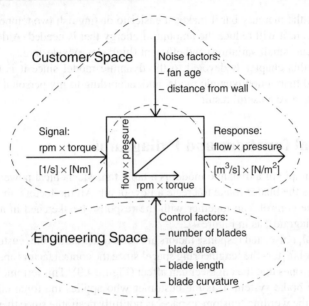

Figure 4.3   The factors of a P diagram are sometimes divided into a customer space and an engineering space. However, this division is not universal since it is not always the case that the signal is out of the control of the engineer.

4.2.1   Noise or signal factor

The pedal force and (depending on the product design) the mechanical power of the table fan are applied by the customer with the intention of changing the response y. This intention is what makes it different from the noise z. The noise might also be caused by the customer, but unintentionally. It should affect the response y as little as possible. The signal M should affect it as much as possible (or possibly aim at a target value, which may be more appropriate for the water tap), so the idea is to find values of the control factors x that in some sense minimizes the sensitivity to noise and maximizes the sensitivity to the signal,

$$\min_x \frac{\partial y}{\partial z}$$

and

$$\max_x \frac{\partial y}{\partial M}.$$

4.2.2   Control or signal factor

Dividing the factors into a customer space and an engineering space as in Figure 4.3 does not always make sense. Sometimes the signal M is a part of the engineering space. The difficult distinction will then become one between the control factors x and the signal M. Let us take an example.

Example 4.1   Sterility is an important property in many industrial applications. It started with for food, where autoclaving has been used since the invention of canned food in the early 19th century, and for medical devices. Nowadays sterilization is an important issue in a much wider range of fields. Some areas of electronics are, for example, electronics for clean laboratories and for space applications (when searching for life at Mars, we cannot bring life from earth). Sterilization can be accomplished in several ways, such as through increased temperature (boiling in water), radiation, or the addition of chemicals. The chemical can be added in wet or dry form (bathing or gassing) and can typically be some kind of alcohol or hydrogen peroxide, H_2O_2. Bandages, for example are sterilized using dry methods and the gases ethylene oxide and formaldehyde are examples of sterilizing agents. The higher the dosage of the chemical, the higher the proportion of inactivated bugs. However, these chemicals are expensive, and the smaller the amount that can be used to achieve a certain sterilization efficiency the better. The slope β in the relation $y = \beta M$ should be as steep as possible. Therefore, even if the dosage can be decided by the engineer and could be regarded as a control factor, it can be economically beneficial to consider it as a signal (Figure 4.4).

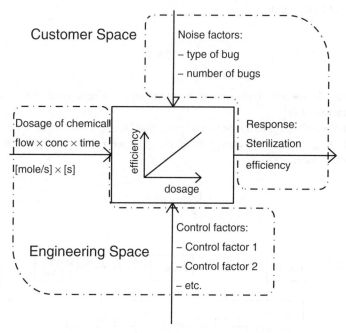

Figure 4.4   A P diagram for the sterilization of Example 4.1.

The conclusion is that if we have some expensive attribute such as energy, matter, time, or force, we will want to use it as efficiently as possible, and therefore it will be worth treating it as a signal rather than a control factor.

4.2.3   The scope

Robust design is accomplished at a detailed level. Scoping down to the simplest act of physics is good advice. The prospects to succeed if, say, an entire brake system is taken into account in a single robust design study are minimal. The brake system must be divided into its elements, as in Figure 4.5. For each one of these elements, it might be possible to accomplish a design that is robust. In order to make the entire brake system robust, each and every one of the elements must be robust. This can either be accomplished by working forward from the force on the pedal or backwards by starting with the friction between the tyre and the road, but will not be discussed here.

Taguchi took this one step further by stating that robust design is best accomplished if the problem is scoped down to one energy transformation. There certainly is a point in this, but he probably overestimated the ability and possibility of the engineer to express every problem in terms of energy transformations.

Example 4.2   A bolt and a nut. The intention of a bolt and nut fastener is to keep things together. The energy used for this can be calculated by integrating the torque with respect to the angle,

$$\int_0^\theta T(\phi)\mathrm{d}\phi,$$

but for the sake of simplicity we will assume that the torque is constant so that the applied energy is $T\theta$.

When the bolt is applied it acts as a spring and Hooke's law can be used. We want the energy to go into spring energy, to an elongation d of the bolt (and compression of the bolted material) as

$$\frac{kd^2}{2} = \beta T\theta,$$

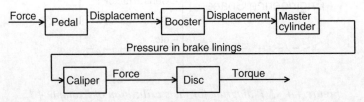

Figure 4.5   Scope down the robust design study to the simplest act of physics.

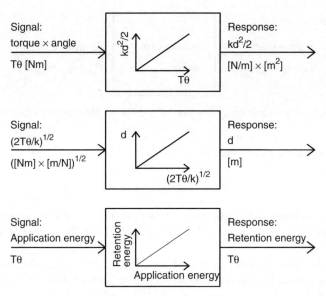

Figure 4.6   Some possible ideal functions for the bolt and nut fastener of Example 4.2.

as in the top diagram of Figure 4.6. However, since we measure the elongation d rather than kd^2, reshuffling the equation as

$$d = \beta\sqrt{\frac{2T\theta}{k}}$$

might feel more natural (Figure 4.6, middle diagram). The ideal function is still based on energy thinking, but it is not always possible to measure the elongation d. However, we usually want the bolt to be easy to apply and hard to remove (it should not fall off due to vibrations). Application energy and retention energy might be a good ideal function,

$$T_{off}\theta_{off} = \beta T_{on}\theta_{on}.$$

This ideal function is also based on energy thinking (Figure 4.6, bottom diagram).

Example 4.3   A rechargeable battery. The aim of a battery is to store energy and make it available for consumption. For a rechargeable battery (e.g. a car battery) the energy that can be taken out of the battery should be high as possible – or as close to the input energy as possible. Thus, the signal can be the input energy,

$$U_{in} \times I_{in} \times t_{in},$$

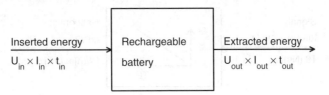

Figure 4.7   A possible ideal function of a rechargeable battery (Example 4.3).

and the response the output energy,

$$U_{out} \times I_{out} \times t_{out},$$

as illustrated in Figure 4.7. Typical noise factors in such a problem can be the storage time (the time between loading and using the energy) and the number of times it has been recharged.

If U_{out} is small, the power from the battery may be so small that the energy is useless. This can be handled in different ways (see Exercise 4.4).

The ideal function for the rechargeable battery of Figure 4.7 is not fully in line with the ideas of Taguchi since it captures more than one energy transformation. The input energy to a rechargeable battery is electrical energy, which is transformed to chemical energy and stored until it is time to be used. Then it is transformed back to electrical energy. Maybe robust design activities would be better if two separate ideal functions had been used, one for loading and one for discharging. However, it might lead to problems in the measurements – the chemical energy is hard to measure. Thus, the ideal function of Figure 4.7 is more appealing than the two of Figure 4.8.

Example 4.4   Some milk and juice packages are equipped with straws that are glued on to the package. The straw should not fall off unintentionally, for example during transportation, but should be easy to remove by the consumer.

The response in such a case can be the energy needed to pull away the straw or the peak force (see Figure 4.9) needed to remove it. There are good arguments for both choices, and these arguments could, for example, be based on a correlation study between the two of them and some subjective customer rating of the effort needed to remove the straw. It is then probably a target value for this (the energy or the peak force).

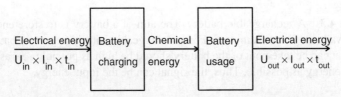

Figure 4.8   Alternative ideal functions for a rechargeable battery.

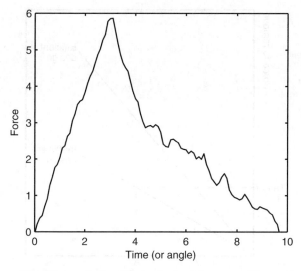

Figure 4.9   The peak force is the peak of the curve (Example 4.4).

Even if this problem can be approached without any signal, it may be better to formulate a dynamic model. Since glue is expensive, we should aim at the target while using as little glue as possible. The amount of glue could then be a signal, but it is hardly a good signal. A large amount of glue does not necessarily mean better attachment. The glued area is another possibility, and it is probably a better one.

However, there is an alternative signal that may be taken into consideration, namely time. The cycle time in the production should be as small as possible. Thus, it is desirable that the response y, no matter whether it is the energy or the force, takes a high enough value even if the holding time (the time when the glue sets and the straw is still unaffected by its own gravity) is short. Apparently, we have at least two different alternatives for signals, area and time, and two alternative responses. A solid subject matter knowledge is needed to pick the best one.

The reason to use a signal in this case is to obtain a product design that is cost efficient. We want a small glued area (or a short time) and despite that be able to reach the target (Figure 4.10). Both time and area could otherwise have been considered as control factors or, at least for the time, not as a factor at all with the motivation that it is given by other project constraints (however, even if it is given, the requirements may be easier to fulfil if it is considered as a signal during the development).

One argument against using the contact area as a signal is that it may be difficult to control. In order to use the contact area as a signal, it might be necessary for this robust design experiment to be preceded by another robust design experiment where the contact area is the response. If this fails and it turns out that the area cannot be made robust, then its nominal value may be used as a control factor and the variation around the nominal as a noise factor in the second experiment.

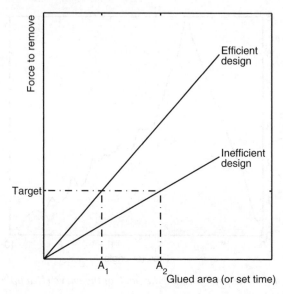

Figure 4.10   Ideal function for the straw removal.

In this application, the direction of the removing force can be a noise factor: no matter in which direction the force is applied it should be equally easy to remove. However, it is not necessarily what we want. If a consumer removes the straw it may be likely that the force is perpendicular to the package. In vibrations or other loads during transportation, for example where the packages are rubbing against each other, the force can be in any direction. Therefore, we might want the force to be dependent upon the direction. In that case, the direction would not be a noise factor.

Example 4.5   A pump. For a pump, there are often a lot of requirements outlining what the pump should not do, such as requirements on audible noise and leakage. However, at least one requirement is usually related to the function. This can, for example, be a requirement on a flow (in m^3/s) given a certain rpm. This relation, rpm to flow, might be good enough to be our ideal function, but it is often better to use energy thinking, shown in the right diagram of Figure 4.11.

There is a difference in view and intent between the two approaches of Figure 4.11. Using the ideal function shown in the left diagram, it is assumed that we get whatever torque we ask for (and get it for free). In the right diagram we have to use the torque we get as efficiently as possible.

If pump A of Figure 4.11 fulfils the requirements (in terms of rpm and flow), but not pump B, which one should be choosen? The answer is not obvious, not even if there is no random variation. If a certain flow is absolutely necessary to have given a certain rpm, the answer is most likely pump A. However, many, maybe even most, requirements are wrong and are set with limited analysis so pump B could very well be a better choice.

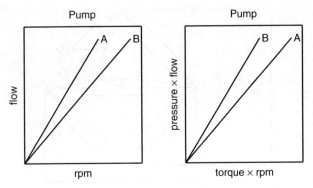

Figure 4.11   If the pump A fulfils the requirements but not pump B, which one should be choosen?

4.3   The signal

It has already been discussed what the signal can be, and three different types of signals were pointed out. One is a signal that the customer applies intentionally in order to change the response. It can, for example, be a force applied to a clutch pedal intending to depress it,

$$y = \beta M + \varepsilon, \quad \varepsilon \text{ is } N(0, \sigma)$$

where y = pedal depression

and M = pedal force.

The other type of signal is more related to control factors and what the engineer can do. One example is sterilization. The higher the dosage of the active chemical, the more spores are inactivated,

$$y = \beta M + \varepsilon, \quad \varepsilon \text{ is } N(0, \sigma)$$

where y = inactivation efficiency

and M = dosage of chemical.

The third situation is a measurement system where it is obvious that there is a signal, but it cannot be classified into any one of the two categories above. For example, consider a Wheatstone bridge (Figure 4.12) with strain gauge sensors used in order to measure pressure. The higher the pressure P, the larger is the change of the voltage ΔV over the bridge. In this case, a high sensitivity factor β is desired. We have

$$y = \beta M + \varepsilon, \quad \varepsilon \text{ is } N(0, \sigma)$$

where $y = \Delta V$

and $M = P$.

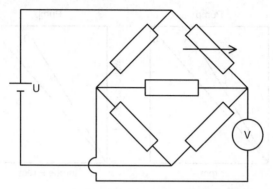

Figure 4.12   For a measurement device like a Wheatstone bridge, the ideal function can be $\Delta V = \beta P$.

4.3.1   Including a signal in a designed experiment

When a signal factor is included in an experimental array, there are at least two things that must be decided. One is the range of the signal. How wide a range should be explored? The other one is the number of signal levels.

When it comes to the range, a reasonable choice can be to span the entire working range and slightly beyond. It is easy to understand what the first part of this statement, the working range, means. For example, the working range of a clutch pedal could be

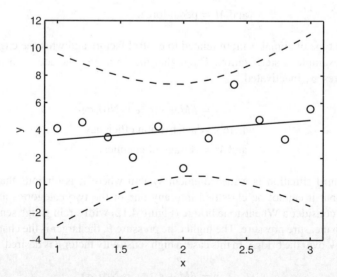

Figure 4.13   Since the uncertainty of a regression line is larger in the ends, exploring the signal effect slightly beyond the range of operation can improve knowledge of the signal effect within this range.

the range from 0 N to the force needed to fully depress the pedal. However, the reason behind the second part of the sentence, 'slightly beyond', is less obvious. There are several reasons. One is that the working range that we can foresee prior to designing the product might be inaccurate. Another one is illustrated in Figure 4.13. If we extend our study slightly beyond the working range, we will gain a better knowledge about the effect of the signal even within the working range.

Our choice of the number of levels will depend on several things. The simplicity of a two-level factorial design is of course a reason to stick to two levels. On the other hand, an assumption of a nonlinear relationship between the signal and response is a reason to make use of several levels. In Chapter 5 we will see that there are different types of experimental array for robust design, combined arrays and crossed arrays. The type of array that is used will affect our choice of the number of levels. For a so called crossed array, the argument to stick to two levels – simplicity – is not relevant. Any number of levels can be used without increasing complexity.

Exercises

4.1 Find ideal functions for some of these:
  (a) An electrical motor.
  (b) A hose clamp.
  (c) A moulding process.
  (d) A strain gauge sensor.
  (e) Something from your own field of work.

4.2 A team is about to design a sterilization mechanism where a chemical in gas form is used to inactivate bugs. They are not certain how to set up the P diagram and need some help. The factors, signals, and responses are:
  • gas concentration;
  • gas temperature;
  • time of exposure;
  • humidity of gas;
  • initial number of bugs, N_0;
  • final number of bugs, N;
  • number of nozzles;
  • preheating of the object to be sterilized;
  • type of bugs.

4.3 A team has decided to make a robust design study on a simple cap of the type used for PET bottles. The team wants to avoid a situation where the consumers find it unpleasant to open, so openability is the issue. The team has noted that they can observe or control a number of things in the test, but would like some help in classifying them as signal, response, noise, or control. These things are:
  • speed of opening, rad/s;
  • total energy;
  • peak force or torque;

- storage time;
- height of grooves;
- radial force from finger to cap;
- roundness of grooves;
- cap diameter;
- cap height;
- cap wet or dry;
- bridges between tampering band and cap, 4 or 8 (same total area);
- contact area, finger–cap;
- cap wall stiffness;
- force (torque) as a function of time (or vice versa);
- force (torque) as a function of angle (or vice versa).

4.4 Recall the rechargeable battery of Example 4.3. The fact that the output voltage may be too small to be useful was left unconsidered. This issue can be handled in different ways.

   (a) Consider the left diagram in Figure 4.14, where the voltage decreases over time. Set up a P diagram with the voltage as the response and the time as a noise factor. What levels of the noise factor 'time' are appropriate?

   (b) Consider the right diagram in Figure 4.14. If it is possible to control the power from the battery in the experimentation, the way to distribute the power over time may be a noise factor. The energy from the battery should be robust against how the power is distributed and it should be possible to take out a high power during a short time or a lower power during a long time. Set up a P diagram with this power distribution as noise.

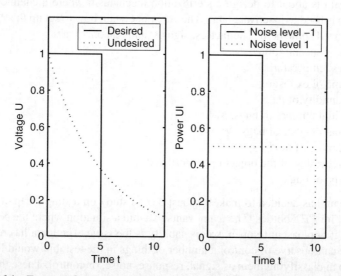

Figure 4.14   For a rechargeable battery, the issue is not just that the energy taken out should be as high as possible, the voltage must also be high enough to make the energy useful (Exercise 4.4).

    (c) Are there any other alternatives for P diagrams for the rechargeable battery?

    (d) Discuss the various P diagrams for the rechargeable battery!

4.5 Consider the problem with the wheel to transport boxes in Example 3.6. Since gravity is used to make the boxes move, there is actually one energy transformation involved.

    (a) Express the problem in terms of a signal and response, for example with the potential energy as the signal and the kinetic energy as the response. How can the fact that the force component working in the direction of box motion changes when the wheel rotation be taken into account?

    (b) Can there be reasons for not expecting a smooth box acceleration and, in that case, how to deal with such a variation over time, for example in terms of factors?

5

DOE for robust design, part 2

5.1 Combined and crossed arrays

5.1.1 Classical DOE versus DOE for robust design

There are some important differences between classical DOE and DOE for robust design. These differences affect both the setup and the analysis of the designed experiment.

- In classical DOE no distinction is made between different types of factors. In robust design, there is a difference between noise factors and control factors.

- In classical DOE it is (mostly) assumed that the variance is one and the same for every combination of control factor levels. In robust design, on the other hand, the idea is to find the setting with the smallest variance.

The first one of these points is not fully true. In classical DOE a distinction is made between ordinary factors and blocking factors. Blocking can actually be regarded as a primitive way to handle noise. However, rather than looking for control-by-noise interactions that can be used to accomplish robustness, the blocking factors are used to obtain better estimates of the effects of the ordinary, 'control', factors.

There are two fundamentally different types of designed experiments for robust design, namely

- combined arrays;
- inner and outer arrays (also called crossed arrays or product arrays).

Inner and outer arrays form the type of design used by the Taguchi School of robust design. Combined arrays are of a later origin and are sometimes ascribed to Welch *et al.* (1990). The type of design treated so far in this book is combined arrays.

Statistical Robust Design: An Industrial Perspective, First Edition. Magnus Arnér.
© 2014 John Wiley & Sons, Ltd. Published 2014 by John Wiley & Sons, Ltd.
Companion website: www.wiley.com/go/robust

Table 5.1   Inner and outer arrays for a robust design experiment.

Control factors				Noise factors			
				$E = -1$ $F = -1$	$E = 1$ $F = -1$	$E = -1$ $F = -1$	$E = 1$ $F = -1$
A	B	C	D				
-1	-1	-1	-1	y_{11}	y_{12}	y_{13}	y_{14}
1	-1	-1	1	y_{21}	y_{22}	y_{23}	y_{24}
-1	1	-1	1	y_{31}	y_{32}	y_{33}	y_{34}
1	1	-1	-1	y_{41}	y_{42}	y_{43}	y_{44}
-1	-1	1	1	y_{51}	y_{52}	y_{53}	y_{54}
1	-1	1	-1	y_{61}	y_{62}	y_{63}	y_{64}
-1	1	1	-1	y_{71}	y_{72}	y_{73}	y_{74}
1	1	1	1	y_{81}	y_{82}	y_{83}	y_{84}

5.1.2   The structure of inner and outer arrays

In a combined array, the noise and control factors are in one and the same array, as shown in Table 2.1. The analysis is primarily made by studying control-by-noise interactions. In inner and outer arrays, two separate arrays are set up that are orthogonal to each other, an inner array (with control factors) and an outer array (with noise factors) in the following way.

Suppose that there are some control factors, say the four factors A, B, C, and D. They go into the inner array. It is a factorial design that is set up for these factors only. For each combination of control factor levels in the inner array there are several runs. However, rather than having replicates or 'a random sample' under each and every combination, as in traditional DOE, it will be a certain structure. The structure is the outer array. The outer array is just another factorial design (Table 5.1). Treating the noise in this manner can be considered as an efficient way to deal with replicates.

For every row, the average and 'standard deviation' are calculated. The standard deviation s is considered as the response. The robustness is maximized if this response s is minimized. The effects that the control factors have on average $\bar{y}$ can be used to move the response to the target value.

Example 5.1   Epitaxial layer thickness (Kackar and Shoemaker, 1986, and Wu and Hamada, 2000). One step in the fabrication of integrated circuits is to grow an epitaxial layer on silicon wafers. The wafers are mounted on a type of six-faceted cylinder, a so called susceptor. Four facets of the susceptor were used in this experiment, two wafers per facet. The susceptor is rotated inside a jar. Chemical vapours are injected into the jar through nozzles at the top of it and heated. The process continues until the epitaxial layer has grown to the desired thickness. In this example, the nominal value for the thickness is 14.5 μm with specification limits of 14.5 ± 0.5 μm. The

Table 5.2   The factors and factor levels of Example 5.1. Data from Wu and Hamada (2000) that cite Kackar and Shoemaker (1986). This material is reproduced with permission of John Wiley & Sons, Inc.

		Level	
Factor	Type of factor	−1	1
A: Susceptor rotation method	Control	Continuous	Oscillating
B: Code of wafers	Control	668G4	678D4
C: Deposition temperature (°C)	Control	1210	1220
D: Deposition time	Control	Short	Long
E: Arsenic flow rate (%)	Control	55	59
F: Hydrochloric acid etch temperature (°C)	Control	1180	1215
G: Hydrochloric acid flow rate (%)	Control	10	14
H: Nozzle position	Control	2	6
L: Location	Noise	Bottom	Top
M: Facet	Noise	1, 2, 3, and 4	

present settings resulted in too large a variation. Thus, the experimenters needed to find process factors that could be set at certain values to minimize the variation of the epitaxial layer thickness, but still with the average thickness as close to the nominal value as possible.

Robust design was considered to be a good way to approach this problem, with the factors and levels of Table 5.2. A crossed array was used. The experimental results are given in Table C.1 in Appendix C.

This setup of a factorial design does not allow an estimation of control-by-noise interactions as for combined arrays. The analysis is performed in another way, namely by studying the variation across each row of Table C.1, that is, for every tested setting of control factors.

For inner and outer arrays, it is possible to explore the data graphically as in Figure 5.1. It is sometimes possible to discern useful patterns in such a way, even without any formal analysis.

In Table 5.3, the averages and standard deviations for each one of the tested control factor settings are given. In the analysis of the data, we will first look for factors having significant effects on $\log \sigma$ but not on the average. A graphical examination of the factor effect plot (Figure 5.2) shows that the nozzle position and the susceptor rotation method can be used to reduce variation and make the process robust against noise. The nozzle should be in position 6 and the susceptor rotation method should be continuous.

The variation can be reduced even further by selecting the deposition time to its low value. However, it seems as if time is the only factor available to tune the system to the target. Therefore, in order to aim correctly, it may be worth the price of a higher variation.

Table 5.3   The averages and standard deviations for each line of the inner array of the wafer data.

A	B	C	D	E	F	G	H	Mean	Standard deviation	log(s)
−1	−1	−1	1	−1	−1	−1	−1	14.79	0.60	−0.51
−1	−1	−1	1	1	1	1	1	14.86	0.14	−1.94
−1	−1	1	−1	−1	−1	1	1	14.00	0.12	−2.10
−1	−1	1	−1	1	1	−1	−1	13.91	0.44	−0.81
−1	1	−1	−1	−1	1	−1	1	14.15	0.07	−2.65
−1	1	−1	−1	1	−1	1	−1	13.80	0.54	−0.62
−1	1	1	1	−1	1	1	−1	14.73	0.68	−0.38
−1	1	1	1	1	−1	−1	1	14.89	0.47	−0.75
1	−1	−1	−1	−1	1	1	−1	13.93	0.83	−0.19
1	−1	−1	−1	1	−1	−1	1	14.09	0.34	−1.09
1	−1	1	1	−1	1	−1	1	14.79	0.54	−0.62
1	−1	1	1	1	−1	1	−1	14.33	0.65	−0.43
1	1	−1	1	−1	−1	1	1	14.77	0.48	−0.74
1	1	−1	1	1	1	−1	−1	14.88	0.81	−0.21
1	1	1	−1	−1	−1	−1	−1	13.76	0.81	−0.21
1	1	1	−1	1	1	1	1	13.97	0.27	−1.32

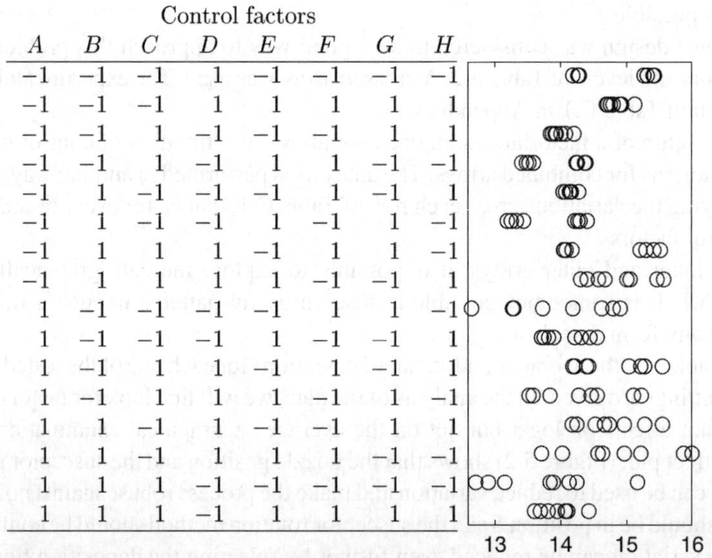

Figure 5.1   A way to illustrate graphically the results of a crossed arrays experiment. Just by looking at the data it may be possible to see if any one of the factors affects the variation and the average.

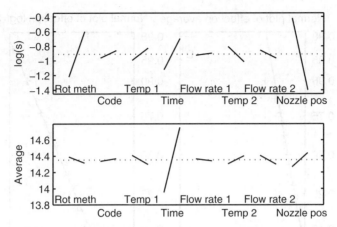

Figure 5.2   In order to reduce the standard deviation, select the nozzle position at the high level and the susceptor rotation method at the low level. The deposition time can be used to tune to target.

A model where the factors pointed out as significant in the normal probability plot (Figure 5.3) are included becomes

$$\log \sigma = -0.91 - 0.49x_H$$
$$\mu = 14.35 + 0.40x_D$$

where

$$x_D = \begin{cases} -1 \text{ if deposition time is short} \\ 1 \text{ if deposition time is long} \end{cases}$$

$$x_H = \begin{cases} -1 \text{ if nozzle position is 2} \\ 1 \text{ if nozzle position is 6} \end{cases}$$

Obviously, nozzle position 6 should be selected to minimize variation (gives $\log \sigma = -1.40$ and $\sigma = 0.25$). In order to select the correct value of the deposition time, we must solve the equation

$$14.5 = 14.35 + 0.40x_D$$

where 14.5 μm is the target value of the thickness. This gives $x_D = 0.37$. Note that this is the time in coded terms. With the numerical values on the deposition time (not just the level descriptions 'short time' and 'long time'), it is possible to calculate the actual time.

The example is worth some reflection. One aspect concerns the noise factors, the four facets of the susceptor and the location. They are both good examples of surrogate

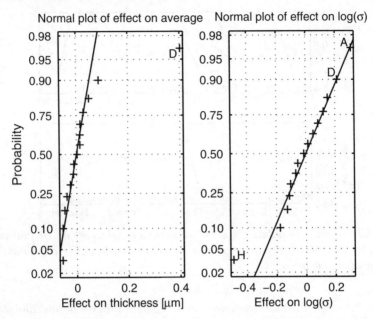

Figure 5.3 Normal probability plots reveal that factor H, the nozzle position, and factor D, the deposition time, are significant effects for log s and ȳ respectively.

noise. However, an even more important reflection concerns the deposition time. In this example, time was treated as a control factor, but there is an alternative approach, namely to use it as a signal. An important aspect of production processes is the production rate. The shorter the cycle time for each unit, the more can be produced. Thus, deposition time is perfectly in line with the discussion in Chapter 4 of what a signal can be. The conclusion is that if our observation in this example that the longer the time the thicker the epitaxial layer is generally valid and not just for the factor levels used in this example, then it can be worth considering deposition time as a signal.

5.2 Including a signal in a designed experiment

5.2.1 Combined arrays with a signal

When signal factors were introduced in this book, it was emphasized that the signal is a factor that is desired to affect the response. A change of temperature shall affect the volume of the mercury bar in a thermometer, a force applied to the clutch pedal shall cause a pedal depression, and an increased chemical dosage shall increase the inactivation efficiency of spores in sterilization. This is totally in contrast to noise factors, for which the effect on the response y should be as small as possible. Thus, for a noise factor z, the control factors $x_1, \dots, x_p$ should be used to minimize the sensitivity,

$$\min_{x_1,\dots,x_p} \left(\frac{\partial g}{\partial z} \right)^2 .$$

For a signal M, on the other hand, this sensitivity should be maximized

$$\max_{x_1,\dots,x_p} \frac{\partial g}{\partial M}$$

or at least aim at a target value β for this slope,

$$\beta(x_1, \dots x_p) = \frac{\partial g(x_1, \dots, x_p, M)}{\partial M} .$$

Example 5.2   Heat exchangers are common devices in both industrial and household applications. In Northern countries like Sweden, they are used in private homes to heat the incoming air in the ventilation system with the outgoing air. In cars, they are used for cooling purposes; air is sucked in through the grill to cool down parts in the engine and turbocharger area. Cooling as in the vehicle application is a much more common usage of heat exchangers in industrial applications than warming up as in the households, since machines and their components typically get warm while in operation and must be cooled down.

This example is based on a real study but differs on a major point from the original. The data are taken from a real case, but the application has been changed for confidentiality reasons. The application was not a heat exchanger but something completely different. Yet we will describe the problem as if it had been a heat exchanger.

In this application, a heat exchanger was used for cooling a liquid. The cooling media was air in a manner resembling one in a car where air is sucked in through the grill and the cooling media runs in channels through the main pipe to cool down rather than in parallel, which is common in many other applications. The signal M is the temperature difference ΔT between the air and the liquid,

$$M = \Delta T = T_1 - T_3$$

and the response of the temperature drop of the liquid (Figure 5.4) is

$$y = T_2 - T_1.$$

The heat exchanger is supposed to work independently of the initial temperature of the liquid, T_1, the humidity of the air, and the age of the heat exchanger. A major concern with the age is corrosion, which can prevent or change the air flow in an undesirable manner. Since humidity may cause corrosion, the heat exchanger may

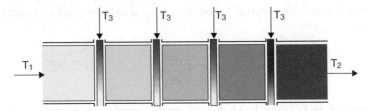

Figure 5.4   Sketch of a heat exchanger of the kind sometimes used in vehicles.

initially work better with high air humidity (the specific heat is higher) but degrade faster. Thus the directionality of the noise factor 'humidity' is not obvious. However, the burden of the experimentation with three noise factors was considered to be too high. After some consideration the humidity was excluded from the experiment and only the initial temperature and the age were kept as noise factors. There were three control factors in the experiment, namely the number of air channels (many narrow or a fewer number but wider), how they are spaced, and one additional factor. The factors are given in Table 5.4.

It was decided that three levels of the signal factor should be used in the experiment. To make use of only two levels on the signal factors would have had some advantages. It would have enabled the usage of the tools for reduced factorial experiments that are available in any software for experimental design. However, the drawback that nonlinear effects are impossible to detect was considered to be so important that three levels were chosen. All other factors have two levels. The experiment was planned as a 2^{5-1} experiment for the remaining five factors (two noise and three control factors), and this was repeated three times, once for each signal level. Such an experimental design may not be the best choice, but it is how it was run. There were three replicates. The signal levels were $M^* = 3$, 4, and 5, but in the analysis these levels have been coded to

$$M = (M^* - 4)$$

so that the coded levels take the values -1, 0, and 1. The experimental data are given in Tables C.2 and C.3.

Table 5.4   The factors of the heat exchanger example.

Factor	Factor type	Levels
A: Number of air channels	Control	Few or many
B: Spacing of air channels	Control	Equidistant or grouped
C: Unspecified	Control	Two levels
D: Initial temperature T_1	Noise	High or low
E: Age	Noise	New or aged
M: Temperature difference ΔT	Signal	Three levels

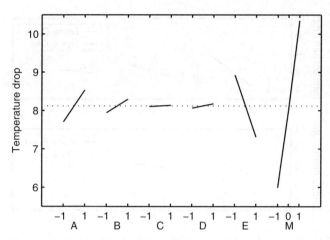

Figure 5.5   Main effects plot for the heat exchanger experiment. Note that the effect of the temperature difference ΔT, here called M, is nearly linear and considerably larger than the other factor effects. This speaks in favour for using it as a signal.

A $3 \times 2^{5-1}$ factorial design with three replicates makes a total number of 144 runs. However, there may be ways to reduce this number. A smaller number of replicates is one way. Sequential experimentation where the first experiment is used to screen out the most important factors is another one, while using two levels rather than three for the signal is yet another one.

Let us first take a graphical look at the effect of the signal in order to judge whether it really can be used as a signal (Figure 5.5). The effect of the signal is steep in comparison to the other factors, which indicates that using it as a signal can be a wise choice. Performing this kind of diagnostic check of the signal is a good habit. It sometimes happens that the factor intended to be the signal cannot be used for the intended purpose, but a graphical check will at least indicate that you are on the right track.

The step to take once the diagnostic check of the signal is ready is to look for control-by-noise interactions that can be used to achieve robustness. It can be performed both graphically (Figure 5.6) and analytically (Table 5.5). The analysis shows that there is one statistically significant control-by-noise interaction, namely the one between the number of air channels and initial temperature. However, no matter if a small or large number of air channels is selected, the response will always be dependent on the initial temperature. The only difference for a low initial temperature is that a large number of air channels will give a higher temperature drop, while for a high initial temperature a small number will give a larger drop. Thus, this interaction cannot be used to obtain robustness.

Figure 5.5 gives a visual picture of the main effects of the control factors. It is often necessary to plot the main effects of the control factors in a separate diagram since their effects may otherwise be overshadowed by the larger signal effect. The scale

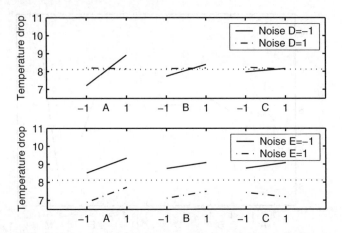

Figure 5.6   Control-by-noise interactions plot for the heat exchanger example.

of the figure will simply be too wide to provide any visual understanding. However, in this application Figure 5.5 is sufficient. Factor *A*, the number of air channels, is the steepest one. An analytical examination shows that it is statistically significant (Table 5.6).

Since it is desired that the function is not only consistent but also efficient, interactions between the control factors and the signal are interesting. These interactions are explored graphically in Figure 5.7 and analytically in Table 5.7. If factor *A*, the number of air channels, is set to the high level, then the response is more sensitive to the signal.

Suppose that a term is included in the model if it is significant at the level $\alpha = 5\%$ using the pooled standard deviation as an estimate of σ. For the heat exchanger, the model becomes

$$y = 8.81 + 0.41x_A + 0.81x_E + 2.18x_M - 0.44x_{AD} + 0.27x_{AM} + \varepsilon.$$

Table 5.5   Control-by-noise interactions for the heat exchanger. The p values are based on the pooled standard deviation for the 48 factor settings, $s = 1.14$.

Interaction	Estimated coefficient	p value
A-by-*D* interaction	−0.44	0.00
B-by-*D* interaction	−0.00	0.99
C-by-*D* interaction	−0.16	0.09
A-by-*E* interaction	−0.01	0.91
B-by-*E* interaction	−0.08	0.40
C-by-*E* interaction	0.14	0.15

Table 5.6   Main effects of the control for the heat exchanger. The *p* values are based on the pooled standard deviation, *s* = 1.14.

Control factor	Estimated coefficient	*p* value
A	0.41	0.00
B	0.18	0.06
C	0.02	0.87

Table 5.7   Main effects of the control for the heat exchanger. The *p* values are based on the pooled standard deviation, *s* = 1.14.

Interaction	Estimated coefficient	*p* value
A-by-M interaction	0.27	0.01
B-by-M interaction	0.07	0.45
C-by-M interaction	0.03	0.72

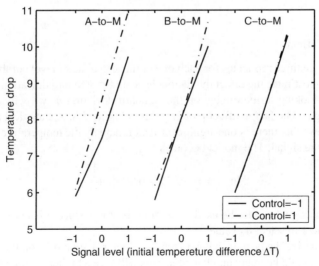

Figure 5.7   Interactions between the signal and the control factors.

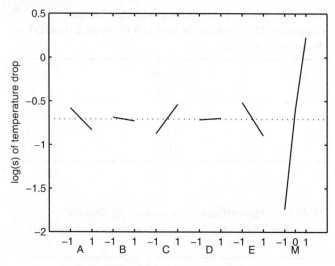

Figure 5.8   Main effect plot for log *s.*

A standard assumption in models of this kind is that the standard deviation σ is independent of the values taken by the factors. However, this assumption is not always fulfilled as there can be dispersion effects as in Section 2.2. There can be settings of the control factors that make the standard deviation smaller. If there are, we have opportunities to make the heat exchanger robust against noise that is not modelled as noise factors but still varies during the course of the experiment.

Dispersion effects

Since there are three replicates for each one of the tested factor level combinations, it is possible to estimate the effect the factors have on the standard deviation. However, rather than looking at the effects on the standard deviation σ, we will look at the effect on log σ. The factor–effect plot is given in Figure 5.8. A statistical analysis of the data shows that there is one significant factor, namely the temperature difference ΔT, that is the signal. The model becomes

$$\log(\sigma) = -0.71 + 0.99 x_M + \varepsilon.$$

However, the signal cannot be used to obtain robustness since it is not in the power of the engineer to select its value.

To summarize the heat exchanger example, factor A should be on the high level, that is there should be many air channels. This will give a higher average both due to the effect of the factor itself and its interaction with the signal. There is an interaction between a control factor, once again factor A, and noise factor D, but it cannot be used to obtain robustness. Both the high level and the low level of factor A give designs that are sensitive to noise factor D, but in different directions (Figure 5.6). It is at least the conclusion if factor A is qualitative and not continuous. If it is continuous, it may at least be worth testing a level in the centre of the interval. None of the two remaining control factors have any significant effect, either as main effects or in interactions.

5.2.2 Inner and outer arrays with a signal

A signal in a crossed array goes into the outer array along with the noise factors. To plan the outer array Taguchi has a 'signal and noise strategy' as one of his steps of robust design. Actually, an explicit step for experimental design is missing in Taguchi's approach. Designing the outer array is namely a part of the 'signal and noise strategy', just as designing the inner array is included in the 'control factor strategy'.

Example 5.3 Airflow from a fan [artificial example]. Consider a small table fan of the type people have at home or place on the desk at their office. The intent of it is to create an air flow (Figure 5.9). The fan is typically driven by an electrical motor and the electrical power is transformed into mechanical power that makes the blades rotate, which in turn moves air. Let us concentrate on the last part, the energy transformation from blade rotation to air movement. The blade power of the rotation is the product of the rpm and the torque and the amount of air moved is the product of the air pressure and the air flow. This gives

$$y = M\beta + \varepsilon,$$

where

$$y = \text{rpm} \times T, \ \left[\frac{1}{\text{s}}\right] \times [\text{N m}]$$

$$M = \text{flow} \times \text{pressure}, \ \left[\frac{\text{m}^3}{\text{s}}\right] \times \left[\frac{\text{N}}{\text{m}^2}\right].$$

Figure 5.9   Air fan.

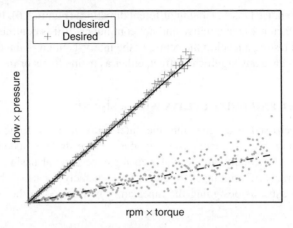

Figure 5.10   In a robust design experiment, we want the relationship between the signal and the response to behave in a consistent and efficient way.

Given a certain signal value, we will not always get the same flow and pressure. Besides the rpm and torque, the output is affected by circumstances such as fan wear, dirt, and disturbing objects located close to the fan. A fan system for which the flow×pressure falls in the steep and narrow cross-marked way of Figure 5.10 is both more consistent (robust) and more efficient than the flatter one.

Assume that the function should be robust against the age of the fan and the closeness to a wall. The closeness is the distance between the back of the fan and the wall. They are both tested at two levels. There are in total eight control factors, and some of them will be tested at three levels, with some others at two levels (Table 5.8).

A full factorial would give an inner array of $2 \times 3^7 = 4374$ combinations. The total number of runs will be much larger since this has to be multiplied by the size

Table 5.8   The factors of Example 5.3.

Factor	Factor type	Number of levels
A: Number of blades	Control	2
B: Blade curvature	Control	3
C: Blade surface	Control	3
D: Blade area	Control	3
E: Blade length	Control	3
F: Centre of gravity	Control	3
G: Blade geometry	Control	3
H: Fan height	Control	3
J: Fan age	Noise	2
K: Distance to wall	Noise	2
M: Torque × rpm	Signal	4

Table 5.9   The inner array of the fan example.

Row	A	B	C	D	E	F	G	H
1	1	1	1	1	1	1	1	1
2	1	1	2	2	2	2	2	2
3	1	1	3	3	3	3	3	3
4	1	2	1	1	2	2	3	3
5	1	2	2	2	3	3	1	1
6	1	2	3	3	1	1	2	2
7	1	3	1	2	1	3	2	3
8	1	3	2	3	2	1	3	1
9	1	3	3	1	3	2	1	2
10	2	1	1	3	3	2	2	1
11	2	1	2	1	1	3	3	2
12	2	1	3	2	2	1	1	3
13	2	2	1	2	3	1	3	2
14	2	2	2	3	1	2	1	3
15	2	2	3	1	2	3	2	1
16	2	3	1	3	2	3	1	2
17	2	3	2	1	3	1	2	3
18	2	3	3	2	1	2	3	1

of the outer array. This makes it necessary to use a fractional design for the inner array. The one used in this example is a type of experimental design called Taguchi's orthogonal L_{18} array. Such an inner array has some drawbacks and some advantages. The major drawback is that it does not allow an estimation of control-by-control interaction effects (discussed further in Chapter 11). The design of the inner array is found in Table 5.9.

The inner array is combined with the outer array. In the outer array, there are two levels for each one of the noise factors and four levels of the signal M, namely the levels 1, 2, 4, and 8. A full factorial array is used for the outer array, is given in which Table 5.10. The experimental results are presented in Tables C.4 and 5.11.

Just as in Figure 5.1, the data can be analysed graphically. Row 1 will then look as in the left diagram of Figure 5.11. Among the 'best' and 'worst' rows in Table C.4 we find rows 8 and 11 respectively (right diagram of Figure 5.11). Note that a graph

Table 5.10   The outer array for the fan example.

Signal	1	1	1	1	2	2	2	2	4	4	4	4	8	8	8	8
J	1	1	2	2	1	1	2	2	1	1	2	2	1	1	2	2
K	1	2	1	2	1	2	1	2	1	2	1	2	1	2	1	2

Table 5.11   The estimated slopes and standard deviations of the fan example.

	A	B	C	D	E	F	G	H	$\hat{\beta}$	s	log s
1	1	1	1	1	1	1	1	1	5.72	3.76	1.32
2	1	1	2	2	2	2	2	2	6.21	4.02	1.39
3	1	1	3	3	3	3	3	3	6.32	2.62	0.96
4	1	2	1	1	2	2	3	3	6.23	3.55	1.27
5	1	2	2	2	3	3	1	1	5.63	2.68	0.99
6	1	2	3	3	1	1	2	2	6.40	3.36	1.21
7	1	3	1	2	1	3	2	3	6.96	2.24	0.81
8	1	3	2	3	2	1	3	1	6.24	1.06	0.06
9	1	3	3	1	3	2	1	2	5.41	2.53	0.93
10	2	1	1	3	3	2	2	1	6.72	2.23	0.80
11	2	1	2	1	1	3	3	2	4.02	5.58	1.72
12	2	1	3	2	2	1	1	3	6.19	6.36	1.85
13	2	2	1	2	3	1	3	2	7.13	3.66	1.30
14	2	2	2	3	1	2	1	3	5.83	1.44	0.37
15	2	2	3	1	2	3	2	1	4.34	4.89	1.59
16	2	3	1	3	2	3	1	2	6.98	2.01	0.70
17	2	3	2	1	3	1	2	3	5.89	3.50	1.25
18	2	3	3	2	1	2	3	1	5.37	4.16	1.42

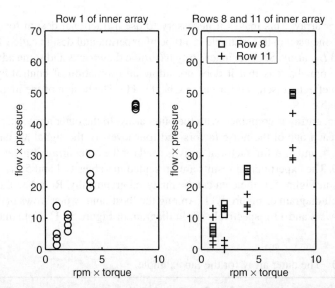

Figure 5.11   The relationship between the signal and the response for rows 1, 8, and 11 of the inner array of the fan experiment. Note that row 8 seems to be better in terms of variation and fan efficiency.

of this kind will also give some insight into the appropriateness of the signal. Can the signal really be used to change the response? In this application it seems to be the case since there is an evident slope.

A glance at Figure 5.11 shows that the combination of control factor levels of row 8 is more efficient than those of rows 1 and 11 since it has a higher slope. It also seems to have a smaller spread, even if it is less obvious, and is thus also more robust.

The efficiency

A visual inspection in this manner is mostly insufficient. In order to analyse the results properly we need to summarize the efficiency and the robustness in some simple numbers. For the efficiency, there are two possibilities that are both appealing and feel natural. One is the slope θ in the regression equation

$$y = \alpha + \theta M + \varepsilon$$

and the other one is the slope β when the regression line is forced through the origin,

$$y = \beta M + \varepsilon.$$

They can both be estimated using the least squares method. The estimates become

$$\hat{\theta}_i = \frac{\sum_{j=1}^{16} \left(M_j - \bar{M}\right) \left(y_{ij} - \bar{y}_{i.}\right)}{\sum_{j=1}^{16} \left(M_j - \bar{M}\right)^2}$$

$$\hat{\beta}_i = \frac{\sum_{j=1}^{16} M_j y_{ij}}{\sum_{j=1}^{16} M_j^2}$$

where M_j is the signal of the jth position in the outer array, $\bar{M}$ is the average signal level, y_{ij} is the jth response at the ith row and finally $\bar{y}_{i.}$ is the average response of the ith row.

A reason to force the regression equation through the origin in this example could be that without any rotation of the blades, the fan will not cause any flow of air. This is perfectly in line with Taguchi's energy arguments. Thus, it is not a coincidence that Taguchi's focus is more on the second one of the models, the one where the regression line is forced through the origin.

The robustness

It is less obvious how to represent the robustness in a single number than it is for the efficiency. One possibility is the standard deviation σ for each fitted regression line (one for each row of the inner array). The logarithm of this quantity, $\log \sigma$, is another possibility. Yet one possibility is some quantity related to the relative standard deviation,

$$\frac{\sigma}{\mu}.$$

However, since we deal with a regression line there is no common expectation μ. A possibility then is to exchange μ for the slope β,

$$\frac{\sigma}{\beta}.$$

Taking the logarithm of this, using the same arguments as for taking the logarithm of σ, would be a further development. We are then very close to the quantity that Taguchi makes use of, namely

$$\eta = 20 \log_{10}\left(\frac{\beta}{\sigma}\right),$$

which is called Taguchi's dynamic signal-to-noise ratio.

The analysis

In the fan data example, we will maximize the slope β and minimize the logarithm of the standard deviation $\log \sigma$ (Table 5.11).

A factor–effect plot for the data of Table 5.11 is given in Figure 5.12. When this plot is interpreted and we try to use the information it provides to learn how to design the fan, the effects can be studied in this order:

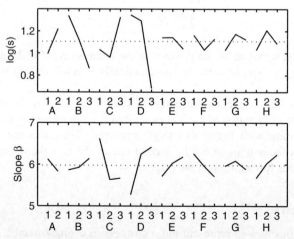

Figure 5.12   For robustness, select $B = 3$, $C = 2$, $D = 3$, and possibly also $A = 1$.

(i) Look for factors that have an effect on the robustness $\log(\sigma)$ (or alternatively σ/β or some other measure of variation) but no effect on the slope. Factor B, the blade area, belongs to this group. To maximize the robustness, select B on level 3. It is possible that factor A belongs to this group as well. In that case, select it to level 1.

(ii) As a second step, look for factors affecting the slope but not the robustness $\log(\sigma)$. Since the slope should be as high as possible in this application (high efficiency), we will select the factors on the levels maximizing the slope. In this example, it does not seem to be any factor of this kind.

(iii) The third step is to look for factors affecting both the robustness and the slope. In this example, factors C, D, and possibly others are of this kind. If a trade-off between robustness and efficiency is needed, it is usually better to sacrifice efficiency for robustness. There are two reasons for that. One is that factors that affect the robustness are rare and those that can be used for this purpose must not be wasted. The other reason is that robustness in most cases is more important than efficiency. Unless a function is robust enough and its response to a signal can be foreseeable, it is useless. In this case, a slightly smaller variation may be obtained if factor C is selected to level 2 rather than level 1, but with a cost of less efficiency.

(iv) Finally, look for factors affecting none of the two, neither the robustness nor the efficiency. Such factors are good since they can be used for something else, for example improving some other property or for cost reduction.

Some issues are worth commenting on. One is factor B. When we select factors to include in the experiment we use our engineering experience. However, it is much easier to have a feeling for the slope rather than the robustness. Therefore, we have a tendency to include factors affecting the slope in the experiment, but dismiss factors like B, the ones affecting the robustness but not the slope, since we judge them as unimportant even before the experiment has started. In this case, we were lucky since factor B is included.

For factor C, the level that is best for the robustness is not the best for the efficiency. The difference in robustness between levels 1 and 2 of factor C is fairly small. With another measure, the robustness of the variation, another conclusion may have been drawn. The relative variation σ/β or Taguchi's SN ratio $20\log_{10}(\beta/\sigma)$ might be more appropriate.

Prediction and confirmation

Just as for the combined array, a graphical analysis should be combined with an analytical one. Assume that such an analysis indicates that factors A, B, C, and D affect $\log(\sigma)$, the robustness, and factors C and D the slope β. If they are selected at their optimal levels with respect to robustness,

$$A = 1$$
$$B = 3$$
$$C = 2$$
$$D = 3$$

we obtain

$$\log \sigma = 1.11 + (0.99 - 1.11) + (0.86 - 1.11) + (0.96 - 1.11) + (0.68 - 1.11).$$

The value 1.11 is the total average and 0.99 the average of $\log s$ when factor A is on level 1, so $0.99 - 1.11 = -0.12$ is the effect of having factor A set to level 1. Since factors like E are insignificant, they are not included in the prediction model. Thus, the prediction of $\log \sigma$ under the optimal setting becomes

$$\log \sigma = 0.16 \Rightarrow \sigma = 1.17.$$

In the same manner, the prediction for the slope β is

$$\beta = 5.98 + (5.64 - 5.98) + (6.41 - 5.98) = 6.07.$$

Once an experiment is performed and an optimal setting is identified, a confirmation run is necessary. Is the prediction model trustworthy and is the result really as good as this prediction points out? For the confirmation, yet another run is performed, now using the optimal settings. The outer array remains the same, so that the new run is perfectly comparable to the ones already performed.

Since it takes a while from the original runs to the confirmation run, some things in the circumstances and environments of the test might have changed. In order to track such a change, it is a good habit to repeat one of the original experiments once the confirmation run takes place. For the fan example, this will mean a total number of 20 rows of the inner array, namely the 18 original and the 2 confirmation runs.

If the experiment does not provide confirmation, there is of course a reason for that. Possible reasons include:

- Control-by-control factor interactions. The inner array of the table fan experiment is extremely reduced and it is not possible to estimate the effect of any control-by-control interaction. However, even if they cannot be estimated, it is possible that they exist.

- Unconsidered noise factors. If an important noise is not considered as a factor in the experiment but varies in an uncontrolled and unobserved way during the runs, then this will affect the response.

If the confirmation run does not give confirmation, a decision must be taken on how to proceed. In fact, there are two possibilities. One is to go back to the planning of the experiment and try to understand which of the control factor interactions that might matter and which noise factors have been left aside and include them in a new experiment. Unfortunately, such an approach is often regarded as too time consuming or too costly. The other possibility is to take the 'best' one of the tested combinations and go along with that one.

5.3   Crossed arrays versus combined arrays

In the choice between combined arrays and crossed arrays, the choice mainly falls on combined arrays (at least if there is no signal). The major reason is the size of the experiment. As will be seen in Exercise 5.3, combined arrays typically require a smaller number of runs to obtain a certain amount of information. However, there are other issues that need to be taken into consideration in the choice of experimental array.

Some arguments in favour of crossed arrays are

- easy to interpret;

- may lead to a limited number of setups in the laboratory (at best one for each row);

- easy to use and interpret if there is a signal factor;

- the confirmation run is easy to interpret.

Arguments speaking against crossed arrays include

- limited flexibility on which interactions to study;

- may require many runs.

Some of these issues speak for themselves, but other are worth commenting on. The statement that the crossed arrays may lead to fewer setups in the laboratory is one of them, and is discussed in Example 5.4. Another one is the inclusion of a signal. We have seen that for a combined array, control-by-signal interactions are treated separately. Thus, for combined arrays with a signal, there are three different kinds of effects to consider, namely control factor effects, control-by-noise interactions, and control-by-signal interactions, and in addition to that there is also a separate study of dispersion effects. For crossed arrays there are only two types of effects, namely control factor effects on the variation and control factor effects on the slope. Two things to consider rather than four is of course an advantage.

A closer examination indicates that it is not even the same property of the slope that is optimized for crossed arrays and a combined array. Maximizing the slope β in crossed arrays deals with a rotation of a line passing through the origin. In the combined array the line is moved upwards by selecting the control factors on levels maximizing the average and is then rotated by using control-by-signal interactions.

However, this slope is unlikely to pass through the origin. In some applications, this is perfectly in order. On other occasions it may penalize designs just because they are also good for low signal levels. Take, for example, the table fan of Figure 5.11. At low levels of the signal, row 11 will hardly blow any air at all. It needs to overcome a threshold level before it delivers any air flow. Row 8 does not have this drawback. A regression line that is not forced through the origin will not penalize designs with a threshold value. Thus, exploiting control-to-signal interactions can be misleading. In contrast to combined arrays, crossed arrays do not have this drawback and are often a better choice in the presence of a signal.

The number of runs is of course a very important aspect in the choice of experimental designs. However, more important than the number of runs is to take a grip on the entire burden of experimentation. As mentioned, crossed arrays may sometimes lead to a smaller number of setups in the laboratory.

Example 5.4   Consider the transportation wheel of Example 3.6. Assume that the problem is handled as a nondynamic one with the distance between boxes as the response and the box and shelf being wet or dry as the noise, just as in Example 3.6. Furthermore, there are five control factors. Two of them are related to the protecting shield, namely its shape and position, and the other three are the hub size, a property of the slide plate, and the surface of the racks (Figures 5.13 and 5.14 and Table 5.12). The surface of the rack is either flat or has grooves. Suppose an experiment is set up in accordance with Table 5.13. Suppose further that the runs are performed in the order they appear in the matrix, row by row. Changing control factor settings may take a while since it includes the change or adjustment of the shield, the slide plate, and the racks. Changing noise factors is easier. It only includes the drying and wetting of the shelves and boxes. This is fairly common, where the change of noise factor levels

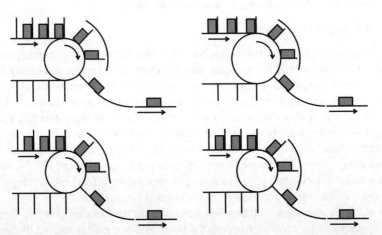

Figure 5.13 Two control factors of Example 5.4 are related to the protecting shield, also called the guide. Top left: baseline, top right: the hub size, bottom left: the shield end shape, bottom right: the angular position.

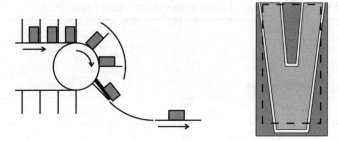

Figure 5.14   In Example 5.4, the box size (dashed line in the right sketch) is larger than the rack. There is a slide plate preventing the box from following the wheel any further. Its angle is control factor E. Left: the position of the slide plate is marked. Right: the rack in light grey, the slide plate in dark grey.

is accomplished easily but the control factor changes are more demanding. Thus, in this case a crossed array may be less cumbersome to use than a combined array.

5.3.1   Differences in factor aliasing

When a fractional factorial design rather than a full factorial is used, something will be lost. Some factors or factor interactions will be confounded or aliased with other interactions, making it impossible to know, given that there is a significant effect, which factor or interaction it is that has this effect. However, this loss of information is often acceptable since higher order interactions are unlikely to exist and the knowledge that would be possible to obtain in a full factorial experiment can be sacrificed for the sake of fewer runs.

Fractional designs are widely used in practice, no matter whether it is in order to learn how robustness can be obtained or for some other reason. However, the aliasing patterns are different if fractionation is used for combined arrays or for crossed arrays. Let us consider an experiment with four factors, three control factors A, B, and C and one noise factor D. In a combined array, the half fractional design in Table 5.14 has

Table 5.12   Control factors for the transportation wheel.

Control factors	
A:	Hub radius
B:	Angular position of shield
C:	Shield end shape
D:	Slide plate
E:	Rack surface

Table 5.13   Possible design of the transportation wheel experiment.

					Noise factor F			
					Wet		Dry	
	Control factors				Replicate			
A	B	C	D	E	1	2	1	2
-1	-1	-1	-1	-1	y_{11}	y_{22}	y_{13}	y_{14}
1	-1	-1	-1	1	y_{21}	y_{22}	y_{23}	y_{24}
-1	1	-1	-1	1	y_{31}	y_{32}	y_{33}	y_{34}
1	1	-1	-1	-1	y_{41}	y_{42}	y_{43}	y_{44}
-1	-1	1	-1	1	y_{51}	y_{52}	y_{53}	y_{54}
1	-1	1	-1	-1	y_{61}	y_{62}	y_{63}	y_{64}
-1	1	1	-1	-1	y_{71}	y_{72}	y_{73}	y_{74}
1	1	1	-1	1	y_{81}	y_{82}	y_{83}	y_{84}
-1	-1	-1	1	1	y_{91}	y_{92}	y_{93}	y_{94}
1	-1	-1	1	-1	$y_{10,1}$	$y_{10,2}$	$y_{10,3}$	$y_{10,4}$
-1	1	-1	1	-1	$y_{11,1}$	$y_{11,2}$	$y_{11,3}$	$y_{11,4}$
1	1	-1	1	1	$y_{12,1}$	$y_{12,2}$	$y_{12,3}$	$y_{12,4}$
-1	-1	1	1	-1	$y_{13,1}$	$y_{13,2}$	$y_{13,3}$	$y_{13,4}$
1	-1	1	1	1	$y_{14,1}$	$y_{14,2}$	$y_{14,3}$	$y_{14,4}$
-1	1	1	1	1	$y_{15,1}$	$y_{15,2}$	$y_{15,3}$	$y_{15,4}$
1	1	1	1	-1	$y_{16,1}$	$y_{16,2}$	$y_{16,3}$	$y_{16,4}$

Table 5.14   A 2^{4-1} design.

	Control		Noise
A	B	C	D
-1	-1	-1	-1
1	-1	-1	1
-1	1	-1	1
1	1	-1	-1
-1	-1	1	1
1	-1	1	-1
-1	1	1	-1
1	1	1	1

Table 5.15   The runs of a 2^{4-1} design are marked 'o', the runs of a crossed array, $2^{3-1} \times 2^1$, with '×'.

Control			Noise	
A	B	C	D = -1	D = 1
-1	-1	-1	o×	×
1	-1	-1		o
-1	1	-1		o
1	1	-1	o×	×
-1	-1	1		o
1	-1	1	o×	×
-1	1	1	o×	×
1	1	1		o

resolution *IV*, meaning that two factor interactions are confounded with other two factor interactions, and the main effects are confounded with three factor interactions. The confounding pattern is

$$I = ABCD, \qquad A = BCD, \qquad B = ACD, \qquad C = ABD,$$
$$D = ABC, \qquad AB = CD, \qquad AC = BD, \qquad AD = BC.$$

A crossed array with the inner array of the type 2^{3-1} and the outer array 2^1 will also require eight runs, just as the combined array. However, it is not the same eight runs in the two experiments (Table 5.15). This will also affect the confounding pattern. It becomes

$$I = ABC, \qquad A = BC, \qquad B = AC, \qquad C = AB, \qquad D = ABCD,$$
$$AD = BCD, \qquad BD = ACD, \qquad CD = ABD.$$

The resolution is only *III*, so from the point of view of traditional DOE, the combined array is better since the resolution is higher. However, in traditional DOE, it is the main effects that are considered as the effects of largest interest, followed by the interactions between two factors as the second most interesting. The reason is that the main effects are most likely to be statistically significant and thus most likely to affect the response. As a consequence, they are the most important effects. In robust design, on the other hand, it is the control-by-noise interactions that are most important, not the main effects. The control-by-noise interactions (factor *D* is the noise) of the crossed arrays are confounded with three factor interactions. Thus, the crossed arrays may have a confounding pattern more appropriate for robust design than a combined array. Bingham and Sitter (2003) have an insightful discussion on this topic.

To be able to make use of all the advantages that the aliasing pattern of crossed arrays may provide requires the data analysis to be performed in the traditional way with just one array, even if the experiment is planned as a crossed array. Only in that way is it possible to obtain detailed knowledge about control-by-noise interactions. If the data are analysed as a crossed array, for example by calculating $\log s$ for each row of the inner array, the only knowledge that can be obtained is that a factor selected at a certain level can accomplish robustness against noise in general, not the detailed information about which noise factor it interacts with. However, with the combined array with eight runs no conclusions at all can be drawn about how to design for robustness.

5.4   Crossed arrays and split-plot designs

5.4.1   Limits of randomization

In a well performed experiment, the order of the runs is randomized. That is a way of avoiding the fact that uncontrolled and unconsidered changes over time affect the conclusions. These changes can be anything from a drift in a measurement system to variation an supposedly homogeneous raw material. The randomization makes the conclusions more trustworthy.

In robust design, randomization is not always possible. It is not just the cost or time of randomization that is a hindrance. It may even be physically impossible. The problem is the so-called surrogate noise. Recall that surrogate noise is a variation over time, between positions or between directions. If it is a variation over time, such as the brake torque in Example 3.4, randomization is not an option since the order of time cannot be manipulated. The variation between different positions, as in the tile baking of Example 3.5, will also make randomization impossible. There is not even an order between the 'runs' since several observations are obtained at one and the same moment, namely every observation from one and the same stack of tiles.

A fundamental topic in statistics is independence between observations. This is actually the reason for randomization. It is known that observations that are close in time to each other can be dependent. To make it more likely that this dependence will have a limited impact on the conclusions, the runs are randomized. It is unfortunate but inevitable that randomization cannot always be used for surrogate noise. Nevertheless, surrogate noise needs to be taken into consideration to obtain robustness and we need to live with the problem.

Another aspect of randomization is seen in Example 5.4 where an inner and outer array is used since it is easier to change the noise factors than the control factors. However, in order to utilize this potential benefit, randomization must be sacrificed and all the runs for a row of the inner array must be run immediately after each other. The internal order within a row can be randomized, and so can the different rows of the inner array, but it is impossible to fully randomize. This kind of experiment with such restrictions on randomization is not at all unknown in statistics and engineering.

It is, as a matter of fact, very common when hard-to-change factors are present and is called a split-plot design.

5.4.2   Split-plot designs

When there is at least one hard-to-change factor and one or several factors that are easier to change, a split-plot design can be a good choice. For example, it can be a moulding experiment with the supplier of granules of plastic (A) as one factor and the holding time (B) as another factor with, say, the shrinkage as the response. In plastic moulding, granules of plastic are fed into a hopper, melted and pressed through an extruder and finally pressed into the cavity. All plastic parts have a tendency to shrink and become slightly smaller after moulding. The more melt that is pressed into the cavity, the higher the back pressure will be. Therefore, the holding time may affect the shrink. Plastic granules from different suppliers may have a slightly different chemical content, which may affect the shrink. However, in order to change granules, the entire system with hopper, extruder, hoses and so on has to be emptied to make certain that there is nothing left. Thus, the granule type is a typical hard-to-change factor.

Assume that a two-level factorial experiment with two replicates is run for this moulding experiment. Assume that a fully randomized design is used. It may look as in Table 5.16. The equipment (extruder, hoses, etc.) has to be emptied seven times during the experiment, even though most engineers will take a shortcut and only empty the equipment before runs 2, 5, and 8, that is three times, since it is the only occasions when the hopper will be filled with another type of granule. However, such a shortcut can sometimes jeopardize the independence between the runs. In a split-plot design, on the other hand, the equipment will be emptied on three occasions in the 'correct' way, and possibly just twice with the shortcut. The idea is to have one level for the granules and one for the holding time picked at random. In the next run, the granule, which form the hard-to-change factor, remain unchanged but the other factor, the holding time, is changed to its other level. Before the third run, the equipment is emptied and filled with the other type of granule, and a holding time is picked at random. In that way, we can obtain the experimental design of Table 5.17. The benefit of split-plot designs is more obvious if there are more factors, especially if there are more easy-to-change factors. The hard-to-change factors in a split-plot design are called the whole-plot factors, with easy-to-change subplot factors.

Table 5.16   Fully randomized 2^2 design with two blocks.

	Run							
	1	2	3	4	5	6	7	8
A	-1	1	1	1	-1	-1	-1	1
B	-1	-1	1	1	-1	1	1	-1

Table 5.17   Split-plot 2^2 design with two blocks.

	\multicolumn{8}{c}{Run}						
	1	2	3	4	5	6	7	8
A	1	1	−1	−1	−1	−1	1	1
B	−1	1	−1	1	1	−1	−1	1

Split-plot designs come in a very natural way for crossed arrays. We saw it in Example 5.4, the transportation wheel. There are two natural cases in which it comes. One is when the noise factors are hard-to-change factors. It can be the case when the noise is environmental conditions during production, such as humidity or temperature, which may require a considerable time to change. The other case is when the control factors are hard to change. That is what we came across with the transportation wheel of Example 5.4.

The kind of split-plot design available in commercial software like Minitab is normally the one with the circles in Table 5.15. It is perfectly in line with traditional DOE and its focus on maximizing the resolution. However, we have seen that this may not be the best choice in robust design. The cross-marked runs of Table 5.15 can be a better choice due to aliasing of the control-by-noise interactions. In addition, there is another thing speaking in favour of the cross-marked runs. In the experiment with the transportation wheel, the problem lies in dismounting the equipment, no matter what is changed. Once it is dismounted, the difference in changing one or several factor settings is small. This may speak in favour of planning the experiment without using commercial software.

Even if the experiment is planned without the help of commercial software, it is an advantage to use such software when the data are analysed. The reason is that the standard deviation of the factor effect estimates in split-plot designs is more complicated than in an ordinary design and commercial software usually handles this in an appropriate way. In an ordinary designed two-level factorial experiment, all estimates of factor effects have the same standard deviation. This makes it possible to, for example, use a common normal probability plot. In split-plot designs, on the other hand, some estimates are only affected by the variation within subplots; other estimates are affected both by the variance within and between subplots.

Let us take an example with one hard-to-change factor, one easy-to-change factor, and two whole-plot replicates, i.e. the type of experiment we have in Table 5.17. For the data there, we can write

$$y_{ij} = \beta_0 + \beta_A x_A + \beta_B x_B + \beta_{AB} x_{AB} + v_i + \varepsilon_{ij}$$

where the index i denotes the subplot number (for Table 5.17 we have $i = 1, \ldots, 4$) and index j the individual observation within a subplot ($j = 1, 2$). Both v and ε are random terms with mean zero, but where v is common for all observations within a subplot,

ε is individual for each observation. If there are also replicates within a subplot, this will add an other index letter. The estimate of the effect of the hard-to-change factor A is

$$\hat{\beta}_A = \frac{1}{8}\left(-y_{11} - y_{12} + y_{21} + y_{22} - y_{31} - y_{32} + y_{41} + y_{42}\right)$$

if the runs are numbered in standard order. The variance of this estimate is

$$Var\left(\hat{\beta}_A\right) = Var\left(\frac{1}{8}\left(-y_{11} - y_{12} + y_{21} + y_{22} - y_{31} - y_{32} + y_{41} + y_{42}\right)\right)$$

$$= \frac{1}{8^2} Var\left(2\sum_{i=1}^{4} v_i + \sum_{i=1}^{4}\sum_{j=1}^{2} \varepsilon_{ij}\right)$$

$$= \frac{2^2 \times 4}{8^2}\sigma_v^2 + \frac{8}{8^2}\sigma_\varepsilon^2.$$

In the estimate of β_B, on the other hand, we obtain

$$\hat{\beta}_B = \frac{1}{8}\left(-y_{11} + y_{12} - y_{21} + y_{22} - y_{31} + y_{32} - y_{41} + y_{42}\right)$$

and thus

$$Var\left(\hat{\beta}_B\right) = Var\left(\frac{1}{8}(-y_{11} + y_{12} - y_{21} + y_{22} - y_{31} + y_{32} - y_{41} + y_{42})\right)$$

$$= \frac{1}{8^2} Var\left(\sum\sum(v_i - v_i) + \sum\sum \varepsilon_{ij}\right) = \frac{8}{8^2}\sigma_\varepsilon^2.$$

For the estimate of the interaction term, the random term v will cancel out in the same way as in the estimate of the easy-to-change effect and will thus have the smaller variance.

In general, we obtain the variance

$$\frac{1}{2^{n_1 - k_1} \times r}\sigma_v^2 + \frac{1}{2^{n_1 - k_1} \times 2^{n_2 - k_2} r}\sigma_\varepsilon^2$$

for the hard-to-change factors and interactions between hard-to-change factors. Here $2^{n_1 - k_1}$ is the size of the array for the hard-to-change factors, $2^{n_2 - k_2}$ the size of the array for the easy-to-change factors, and r the number of whole plot replicates. For all other factor estimates, the variance is

$$\frac{1}{2^{n_1 - k_1} \times 2^{n_2 - k_2} r}\sigma_\varepsilon^2.$$

Split-plot designs in robust engineering are discussed further in, for example, Bingham and Sitter (2003).

Exercises

5.1 Analyse the wafer data of Table C.1 not just graphically but also analytically. Use the analytical method you are used to and prefer (e.g. the PRESS value, the predictive R^2, or the p values from F tests and any appropriate method, e.g. stepwise regression, in the selection of terms to include). Will the conclusions differ from those of the normal probability plot (Figure 5.3)?

5.2 A milk package is usually, formed by making a sleeve tube of a carton sheet, sealing it along the tube (the longitudinal sealing) and in one end, and then after filling also in the other end, the so-called transversal sealing. There may then, at least for some types of milk packages, be a misalignment called a hammock in the transversal sealing (Figure 5.15). There is a target value for this, namely zero. The hammock can be in either direction, so the values can be positive or negative. Noise that against which the hammock should be robust include the position along the transversal sealing and whether the package is rotated, that is if the longitudinal sealing is in the centre of one side panel (as it should be) or not. A team of engineers performed an inner and outer array type DOE for this (Table C.5). Analyse the data.

5.3 Consider a problem with four control factors and three noise factors. Each factor has two levels.

    (a) In crossed arrays, how many runs are needed to avoid two-factor interactions to be confounded with main effects or another two-factor interaction (to get a design of resolution V or higher)?

    (b) How many runs are needed in a combined array under the same requirement?

    (c) How many runs are needed in a combined array to avoid the situation where control-by-noise interactions are confounded with main effects or other two-factor interactions?

    (d) In crossed arrays, how does the confounding pattern look if the inner array is of the type 2^{4-1} and the outer array 2^3? [Hint: write all runs in a common array to simplify the analysis.]

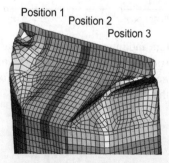

Figure 5.15   A hammock is a misalignment of the two sides in a transversal sealing of a package.

(e) Compare the crossed arrays in (d) with an ordinary half fractional combined array, 2^{7-1}, in the same way as in Table 5.15.

(f) In (d), the inner array is fractional but the outer one is not. Would there be any advantages or drawbacks to fractionate the outer array but not the inner array?

5.4 Set up an inner and outer array with four factors on two levels each, three control factors A, B, and C, and one noise factor D. Suppose that the true model is

$$y = 12.7 + 3x_A + 2x_D - 1x_{AD} + \varepsilon, \qquad \varepsilon \sim N(0, \sigma)$$

where $\sigma = 1$. Generate random data from this model with an optional number of replicates. Analyse the result and compare the obtained estimate of σ with the value in the model. Reflect on the result.

References

Bingham D and Sitter RR (2003) Fractional factorial split-plot designs for robust parameter design. *Technometrics*, **45**, 80–89.

Kackar RN and Shoemaker AC (1986) Robust design: a cost-effective method for improving manufacturing processes. *AT & T Technical Journal*, **65**, 39–50.

Welch WJ, Yu TK, Kang SM, and Sacks J (1990) Computer experiments for quality control by parameter design. *Journal of Quality Technology*, **22**, 15–22.

Wu J and Hamada M (2000) *Experiments. Planning, analysis, and parameter design optimization.* John Wiley & Sons, Inc.

6

Smaller-the-better and larger-the-better

6.1 Different types of responses

Let us recall a few examples that we have come across so far in this book, such as Example 2.1. The target is to have a strip position that is robust against the temperature of the board. Another is Example 5.1, where the epitaxial layer thickness should be robust against the location and facet. There were target values both for the layer thickness, 14.5 μm, and for the strip position. In fact, in each and every example we have come across up to now there has been a target value, at least if the dynamic model is excluded. However, in real life there are other types of problems as well. For example, it could be flaws on a silicon wafer, typically something that should be avoided. Consequently, the number of flaws is a response that should be minimized. This is something different from all the problems and exercises yet considered where the response y can have taken values on either side of the target, smaller or larger. However, the number of flaws cannot be negative. This type of problem is called 'smaller-the-better', or just STB. It is worth noticing that this is completely different from what we came across in Exercise 5.2 where the hammock of a transversal sealing was under study. The target value of the hammock is zero, just as for the number of flaws, but, contrary to the flaw number, the hammock can take negative as well as positive values.

Besides problems of the type smaller-the-better, there are also larger-the-better (LTB) problems. An example is lifetimes. In addition, there is a type of problem that can be considered as a combination of the two. Process windows (or operating windows as they also are called) can be handled using robust design. Typically, a process window is a window for a process parameter in manufacturing, such as

Statistical Robust Design: An Industrial Perspective, First Edition. Magnus Arnér.
© 2014 John Wiley & Sons, Ltd. Published 2014 by John Wiley & Sons, Ltd.
Companion website: www.wiley.com/go/robust

a temperature or pressure value. If the process parameter falls within the process window, the manufacturing process 'works well'. The process window should be as wide as possible. This can sometimes be accomplished by treating the lower bound of the window as a smaller-the-better problem and the upper bound as a larger-the-better one, and addressing the two limits at one and the same time.

It is thus possible to classify robust design problems into different categories, namely

- dynamic models;

- nondynamic nominal-the-best (NTB);

- smaller-the-better (STB);

- larger-the-better (LTB);

- operating window.

The two first on the list have already been discussed, so the rest of this chapter will be devoted to the last three ones. For further reading, Wu and Wu (2000) give a good discussion on these different types of responses and how they can be addressed with inner and outer arrays and analysed with ANOVA.

6.2   Failure modes and smaller-the-better

6.2.1   Failure modes

A fundamental idea of Taguchi is to focus on what the product is intended to do. In that way, he stated, no energy will be left over to cause failure modes or undesired variation, and several failure modes are avoided in one and the same instant shot. The failure modes are just symptoms of nonrobustness. There is certainly a point in this, but this idealistic view is not always practically useful. We sometimes need to deal with failure modes as the response. We will study how this can be accomplished and will see that the approach can be smaller-the-better.

The basic intention of an electrical motor is to transform electrical power to mechanical power. However, commonly there is also considerable heat dissipation. The motor becomes warm. The higher the electrical power – or more possibly electrical energy – the more heat. If this is approached with robust design, Taguchi would probably have suggested the usage of a dynamic model of the type

$$y = \beta M$$

where y = mechanical power, torque $\times$ rpm

and M = electrical power, UI (where U is voltage and I is current).

In this way, the problem is expressed in the form of what the motor is intended to do, and if it succeeds by making it robust 'there will be no energy left over to create heat', as Taguchi might have stated. However, there are other opportunities. Some

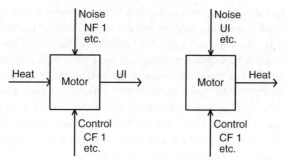

Figure 6.1   Two possible P diagrams for an undesired outcome, heat.

are good and some are not. For example, it is obvious that the approach

$$y = \beta M$$
where y = heat
and M = electrical power, UI

will not work. The reason is that a higher amount of heat when the electrical power UI is increased is undesirable, although it is inevitable. On the contrary, the amount of heat dissipation should be as robust against the electrical power as possible. Thus, the electrical power can be noise, but not a signal. It becomes natural to set up the problem without any signal and express it as a smaller-the-better problem. This is typical for problems of this type; the factor that would have been a signal if the intended function had been studied will be noise if a failure mode is the response.

There is also another approach that can be taken. This approach is not at all logical since the principle of causality is sacrificed. The cause will be used as the response and the effect as the signal. For the heat this means that

$$y = \beta M$$
where y = UI
and M = heat.

Using such a model, the slope β should be maximized and the variation minimized. It is, of course, just a mathematical trick, but is useful and simple. Taken together, this means that for a failure mode of this kind, the two P diagrams of Figure 6.1 are both possible.

6.2.2   STB with inner and outer arrays

So far in this book, the importance of a two-step approach, to first accomplish robustness and then move to the target, has been stressed over and over again. For inner and outer arrays, this means first to reduce the standard deviation σ and then to use the remaining factors to move μ to the target. For a combined array, the first step is to look for control-by-noise interactions and the second step for control factor effects. However, for STB, this two-step approach is not very useful and has a tendency to result in product and process designs that do not work very well. For combined arrays, the remedy is usually to change the order of the two steps; first reduce the average, and then look for control-by-noise interactions to increase robustness. The problem can be treated in a similar manner for inner and outer arrays, first to minimize μ and then to minimize $\log \sigma$. However, for inner and outer arrays there is another alternative as well, namely to make use of the loss function (Appendix A) and in that way minimize average and variation at the same time.

Let

$$m = \text{target value}$$

so that the loss function $L(y)$ becomes

$$L(y) = K (y - m)^2 .$$

Since $m = 0$ for STB, we obtain

$$L(y) = Ky^2$$

and thus the expected loss is

$$E \left[KY^2 \right] = K \left(\sigma^2 + \mu^2 \right) .$$

For a random sample

$$\left(y_1, \ldots, y_n \right)$$

the average loss becomes

$$\frac{K}{n} \sum_{i=1}^{n} y_i^2 = \frac{K}{n} \sum_{i=1}^{n} \left(y_i - \bar{y} + \bar{y} \right)^2$$

$$= \frac{K}{n} \left(\sum_{i=1}^{n} \left(y_i - \bar{y} \right)^2 + \sum_{i=1}^{n} \bar{y}^2 + 2\bar{y} \sum_{i=1}^{n} \left(y_i - \bar{y} \right) \right)$$

$$= K \left(\frac{n-1}{n} s^2 + \bar{y}^2 \right) .$$

It is this average loss, or $\log E[L]$, that is minimized in smaller-the-better problems.

Example 6.1 Consider the data of Tables 6.1 and 6.2. The target is to minimize the response y. If this is done by minimizing the total loss, it turns out that factor A should be at level $A = -1$ (Figure 6.2). It is worth noticing that if the problem is approached using the traditional two step approach, factor A will be selected to have value $A = 1$.

Table 6.1   Smaller-the-better data of Example 6.1.

| Control factors | | | Noise factors and replicates | | | | | | | |
| | | | $D = -1$ | | | $D = 1$ | | | | |
A	B	C	1	2	3	1	2	3	$\bar{y}$	s
−1	−1	−1	1.29	1.30	1.38	0.35	0.42	0.39	0.86	0.52
1	−1	−1	1.39	1.48	1.57	1.59	1.54	1.56	1.52	0.07
−1	1	−1	1.37	1.30	1.42	0.42	0.38	0.43	0.89	0.52
1	1	−1	1.62	1.63	1.60	1.56	1.55	1.53	1.58	0.04
−1	−1	1	1.07	1.09	1.09	0.31	0.35	0.34	0.71	0.41
1	−1	1	1.29	1.25	1.39	1.33	1.39	1.41	1.34	0.06
−1	1	1	1.16	1.19	1.12	0.34	0.32	0.36	0.75	0.45
1	1	1	1.45	1.34	1.40	1.42	1.34	1.44	1.40	0.05

Table 6.2   Loss function data of Example 6.1.

| Control factors | | | | | | | |
A	B	C	$\bar{y}$	s	Total loss	$\sum(y_i - \bar{y})^2$	$\sum \bar{y}^2$
−1	−1	−1	0.86	0.52	5.73	1.33	4.40
1	−1	−1	1.52	0.07	13.90	0.03	13.88
−1	1	−1	0.89	0.52	6.09	1.37	4.72
1	1	−1	1.58	0.04	14.99	0.01	14.98
−1	−1	1	0.71	0.41	3.83	0.84	2.99
1	−1	1	1.34	0.06	10.86	0.02	10.84
−1	1	1	0.75	0.45	4.37	0.99	3.37
1	1	1	1.40	0.05	11.76	0.01	11.75

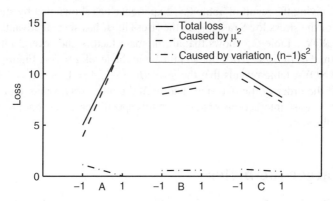

Figure 6.2   Main effects on the total loss and its components in Example 6.1.

Table 6.3   The factors and factor levels of Example 6.2.

Factor	Type of factor	Low level	High level
A: Unspecified	Control	−1	1
B: Unspecified	Control	−1	1
C: Material	Control	Type 1	Type 2
N: Force position and direction	Noise	−1	1

The two-step approach does not work for a reason. In Table 6.2, the loss is divided into its two components, the loss for not aiming at the target and the loss due to variation. The first one of these two terms is considerably larger than the second. This is often the case for smaller-the-better problems and a major reason for abandoning the two-step approach. However, since the result will not be a design that prevents variation to propagate, it is not robust design in a more dogmatic sense, but it works.

6.2.3   STB with combined arrays

As mentioned earlier, the two-step approach of robust design, where we first look for control-by-noise interactions to improve robustness and then for control factors affecting μ to move to target, does not work very well for STB. The order needs to be interchanged, as also recommended by Wu and Hamada (2000).

Example 6.2 Mechanical stress S. In mechanical engineering, the mechanical stress S is an important property. For example, stress variation over time, such as that caused by vibrations, can lead to fatigue and shorten the lifetime of the product. The stress should be as small as possible. An experiment was performed to make the stress robust against the action point and direction of a force. The factors of Table 6.3 were used, giving the result of Table 6.4. Experience has shown that by studying the logarithm of the stress $\log S$ rather than the stress itself has several advantages.

The ANOVA Table 6.5 shows that all the main factors and several interactions, among them AB, BC, and ABC, are significant at the level $\alpha = 5\%$. Figure 6.3 along with the ANOVA table reveals that the levels $B = -1$, $A = 1$, and $C = 1$ should be selected. If the order of the two steps had been the other way round, that is to look at control-by-noise interactions first, then in this specific case the conclusions would have been the same.

6.3   Larger-the-better

Larger-the-better problems are typically what we have when dealing with lifetimes. One way to approach the problem if crossed arrays are used is to rewrite the problem

Table 6.4   The measurement data of Example 6.2.

Control factors			Noise	Response	
A	B	C	N	S	log S
−1	−1	Type 1	−1	9.49	2.25
−1	−1	Type 2	−1	18.17	2.90
−1	1	Type 1	−1	97.31	4.58
−1	1	Type 2	−1	23.97	3.18
1	−1	Type 1	−1	8.51	2.14
1	−1	Type 2	−1	4.55	1.52
1	1	Type 1	−1	373.01	5.92
1	1	Type 2	−1	53.85	3.99
−1	−1	Type 1	−1	10.66	2.37
−1	−1	Type 2	−1	105.37	4.66
−1	1	Type 1	−1	283.47	5.65
−1	1	Type 2	−1	61.12	4.11
1	−1	Type 1	−1	12.36	2.51
1	−1	Type 2	−1	4.93	1.60
1	1	Type 1	−1	77.82	4.35
1	1	Type 2	−1	50.06	3.91
−1	−1	Type 1	1	29.03	3.37
−1	−1	Type 2	1	53.64	3.98
−1	1	Type 1	1	553.16	6.32
−1	1	Type 2	1	82.56	4.41
1	−1	Type 1	1	17.57	2.87
1	−1	Type 2	1	3.97	1.38
1	1	Type 1	1	503.08	6.22
1	1	Type 2	1	75.73	4.33
−1	−1	Type 1	1	30.64	3.42
−1	−1	Type 2	1	141.39	4.95
−1	1	Type 1	1	487.31	6.19
−1	1	Type 2	1	63.97	4.16
1	−1	Type 1	1	12.59	2.53
1	−1	Type 2	1	14.80	2.69
1	1	Type 1	1	679.91	6.52
1	1	Type 2	1	129.43	4.86

into a smaller-the-better form,

$$\min \sum \frac{1}{y_i^2}.$$

Another alternative is (1) to minimize the relative variation, σ/μ, or better the logarithm of it, and (2) to maximize μ.

Table 6.5   The ANOVA table for the stress measurements.

Source	$\hat{\beta}$	SS	d.f.	MS	F	p
A	−0.29	2.62	1	2.62	7.65	0.014
B	1.05	35.20	1	35.20	102.75	0.000
C	−0.33	3.50	1	3.50	10.21	0.006
D = noise	0.39	4.93	1	4.93	14.39	0.002
AB	0.38	4.62	1	4.62	13.49	0.002
AC	−0.22	1.52	1	1.52	4.44	0.051
AD	−0.05	0.08	1	0.08	0.25	0.627
BC	−0.47	7.05	1	7.05	20.58	0.000
BD	0.06	0.14	1	0.14	0.39	0.549
CD	−0.09	0.24	1	0.24	0.71	0.413
ABC	0.28	2.48	1	2.48	7.23	0.016
ABD	0.06	0.13	1	0.13	0.38	0.546
ACD	0.03	0.02	1	0.02	0.06	0.807
BCD	−0.05	0.08	1	0.08	0.23	0.640
ABCD	−0.04	0.04	1	0.04	0.13	0.726
Error		5.48	16	0.34		
Total	3.97	68.13	31			

If the experiment is performed in terms of a combined array, the order of the two steps will be, first, to maximize the average and, then among the remaining factors make use of control-by-noise interactions to improve robustness. This reversion of the steps should not be astonishing. Take lifetimes, for example. A bracket with an average lifetime of one million cycles in some application typically has a larger standard deviation than one with an average lifetime of ten thousand cycles. If the experiment includes replicates, minimizing σ/μ could be a first step (since the term involves the average), just as for the inner and outer arrays. However, utilizing the control-by-noise interactions cannot come before the maximization of μ.

6.4   Operating window

Example 6.3 Copy machine (from Phadke, 1989). Consider the paper feed of a copy machine (Figure 6.4). Primarily two failure modes that are connected to the force can occur. If the force is too small, there will be no paper feed. If it is too high, it will be a multifeed. The target is to maximize the window between no feed and multifeed, the operating window. Let

$$y_i = \text{lowest force to feed paper}$$

and

$$v_i = \text{highest force avoiding multifeed.}$$

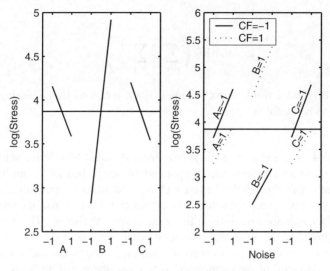

Figure 6.3   Main and interaction plots for log *S.*

In order to maximize the window, there are some possibilities. One is to maximize some form of $V - Y$, for example

$$\max \ \mu_v - \mu_y$$
$$\text{or} \ \max \ P(V - Y) > a.$$

Another alternative is to consider it as a multiobjective problem, where y is treated as a smaller-the-better problem and v as a larger-the-better one. However, the approach we will take here is another one, namely to combine STB and LTB,

$$\min \left(\sum_{i=1}^{n} y_i^2 \ \sum_{i=1}^{n} \frac{1}{v_i^2} \right)$$

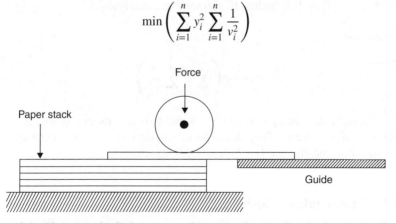

Figure 6.4   The paper feed of a copy machine. Phadke, Madhav S., Quality Engineering Using Robust Design, *1st Edition, © 1989. Reprinted by permission of Pearson Education, Inc., Upper Saddle River, NJ.*

or better

$$\min \log \left(\sum_{i=1}^{n} y_i^2 \sum_{i=1}^{n} \frac{1}{v_i^2} \right).$$

This idea originates from Xerox (Phadke, 1989, refers to D.P. Clausing) and has turned out to be useful.

Example 6.4 Wave soldering (originally from Peace, 1993). Wave soldering is a method used to attach electronic components to circuit boards. It can be used for circuit boards with plated holes and through-hole mounted components. The board is passed over a wave of molten solder. The circuit board is on top of the wave and the molten solder is pushed up through the holes to form the solder. The other parts of the circuit board are covered with a solder mask in order not to be wetted. There are mainly two failure modes in this type of soldering, solder voids and solder bridges. If the top side board of the temperature is too low, there will be voids, and if it is too high, there will be bridges. The temperature window should be maximized. The experimenters had in total 15 control factors to consider. An inner and outer array was used, with an inner array of the type 2^{15-11}. The factors are given in Table 6.6, the experimental array in Table D.1, and the results in Table D.2 (both in Appendix D), which in turn are summarized in Table 6.7. The noise factor is the board carriers. There were five of them, so this is the number of noise factor levels.

Let

$$Y = \text{lowest temperature to avoid voids}$$

and

$$V = \text{highest temperature to avoid bridges}$$

and minimize

$$\min \log \left(\sum_{i=1}^{n} y_i^2 \sum_{i=1}^{n} \frac{1}{v_i^2} \right).$$

A graphical data analysis (Figures 6.5 and 6.6) reveals that factors J, H, and possibly D are significant. They should be selected at levels $J = $ small pad size, $H = $ low solder wave height, and $D = $ hole size 0.039 inch.

6.4.1   The window width

This approach to operating windows does not give the window itself. For the wave soldering example, we learn that factors D, H, and J affect the window width, but not

Table 6.6   The control factors of the wave soldering. From Peace (1993).

	Levels	
Factor	Low	High
A: PCB finish	Hot air level	Bare copper
B: Solder mask photo-imageable	Liquid	Dry film
C: PCB thickness	0.062 inch	0.090 inch
D: Plated through hole size	0.033 inch	0.039 inch
E: Flux air knife pressure	40 psi	60 psi
F: Flux air knife angle	0 degrees	45 degrees
G: Flux wave height	Low	High
H: Solder wave height	Low	High
I: Flux composition	Low solids	Medium solids
J: Pad size	Small	Large
K: Solder pump size	55%	80%
L: Flux density	0.86 g/ml	0.87 g/ml
M: Solder temperature	460°F	500°F
N: Solder waves	Single	Dual
O: PCB orientation	0 degrees	45 degrees

Table 6.7   Some values calculated from the solder wave experiment.

Row	$\sum y_i^2$	$\sum 1/v_i^2$	$\log \sum y_i^2$	$\log \sum 1/v_i^2$	$\log \sum y_i^2 + \log \sum 1/v_i^2$
1	297 223	7.49E-05	12.60	−9.50	3.10
2	268 673	8.95E-05	12.50	−9.32	3.18
3	249 595	6.40E-05	12.43	−9.66	2.77
4	269 154	9.54E-05	12.50	−9.26	3.25
5	276 670	9.26E-05	12.53	−9.29	3.24
6	279 033	7.51E-05	12.54	−9.50	3.04
7	272 874	7.88E-05	12.52	−9.45	3.07
8	275 690	8.76E-05	12.53	−9.34	3.18
9	283 731	6.62E-05	12.56	−9.62	2.93
10	245 259	1.00E-04	12.41	−9.18	3.23
11	234 689	7.09E-05	12.37	−9.55	2.81
12	261 305	9.35E-05	12.47	−9.28	3.20
13	275 696	8.83E-05	12.53	−9.33	3.19
14	277 115	8.65E-05	12.53	−9.36	3.18
15	239 402	7.90E-05	12.39	−9.45	2.94
16	241 609	7.51E-05	12.40	−9.50	2.90

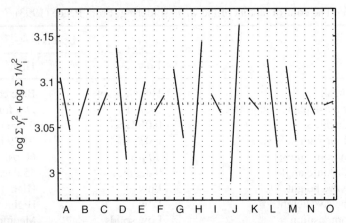

Figure 6.5   Factor–effect plot for the response $\log \sum y_i^2 + \log \sum \left(1/v_i^2\right)$ *of the wave soldering example.*

what the minimum and maximum temperatures of the window are. These have to be calculated separately.

It is not obvious how this window should be calculated. Regression models for both the averages, μ_y and μ_v, and both standard deviations, σ_y and σ_v, are required, or alternatively models for their logarithms. However, which terms to include in the models is less clear. One possibility is to make use of a model with the factors D, H, and J only since these are the ones that turned out to be the factors affecting the

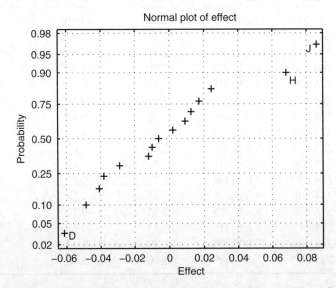

Figure 6.6   Normal probability plot for the wave soldering (Example 6.4).

window width. Another possibility is for each one of the four responses to make use of a subset of these three factors, namely the ones that are statistically significant for that specific response, which will be done in this example (a better alternative is used in Exercise 6.4d).

Since the calculations for finding the significant factors contained logarithms, it seems natural to continue to use logarithms to identify the window. We obtain

$$\log\left(\hat{\mu}_y\right) = 5.4386 - 0.0188 x_D$$
$$\log\left(\hat{\sigma}_y\right) = 1.2008$$
$$\log\left(\hat{\mu}_v\right) = 5.5107 - 0.0351 x_H - 0.0460 x_J$$
$$\log\left(\hat{\sigma}_v\right) = 1.7028.$$

For the values $D = 1$, $H = -1$, and $J = -1$, which are the optimal settings, the numerical values are

$$\log\left(\hat{\mu}_y\right) = 5.4386 - 0.0188 = 5.4198$$
$$\log\left(\hat{\mu}_v\right) = 5.5107 + 0.0351 + 0.0460 = 5.5918$$

and thus

$$\hat{\mu}_y = 225.8, \hat{\sigma}_y = 3.3$$
$$\hat{\mu}_v = 268.2, \hat{\sigma}_v = 5.5.$$

If the temperature $\mu_y + 3\sigma_y$ is considered as 'safe' with respect to the failure mode voids and $\mu_v - 3\sigma_v$ safe with respect to bridges, then the window becomes

$$[235.7 \text{ degrees}; \ 251.7 \text{ degrees}].$$

Exercises

6.1 Consider the data of Example 6.1.
  (a) Identify the significant effects and build a statistical model.
  (b) Analyse the data in terms of the logarithm of the loss, $\log L$. Will the conclusions be the same?
  (c) Since both the inner and outer arrays are full factorial designs, the data can be reorganized into a combined array. Analyse the data in this form.

6.2 Consider the stress data of Example 6.2.
  (a) Since there are two replicates, it is possible to estimate the effects each factor has on $\log \sigma$. Perform this analysis.
  (b) How can the product be designed after consideration is taken of all relevant and significant factors, both the effects on μ (main effects and interactions)

Table 6.8   The factors and factor levels of Exercise 6.3.

	Type of factor	Low level	High level
A: Ink type	Control	Type 1	Type 2
B: Ink temperature	Control	Low	High
C: Unspecified	Control	Low	High
D: Surface roughness	Noise	Low	High
E: Paper humidity	Noise	Dry	Wet

and the effects on $\log \sigma$? Make use of relevant criteria for deciding which factors to include and which to exclude.

(c) Analyse the data in their original form, without taking the logarithm. Reflect on the approach.

6.3 White spots. A paper, a postcard, or a poster that is printed is never perfectly flat. There are cavities. When printed, the ink cannot cover the cavities. It causes white spots, a failure mode that is undesirable for posters. The problem can be analysed by painting the surface with a single colour and using image analysis. The response is the number of white spots (an alternative is the white area). This number should be minimized. A robust design experiment using the factors of Table 6.8 was carried out, giving the results of Table 6.9. The different levels of the paper humidity were controlled in the experiment by storing the papers in climate chambers during a long enough time before the ink application took place. Analyse the data.

6.4 Consider the wave soldering problem in Example 6.4.

(a) Analyse the data by separately minimizing $\log \sigma_y$, $\log \sigma_v$, and $\log \mu_y$ and maximizing $\log \mu_v$. Are there any contradictory conclusions on the optimal factor settings?

(b) Analyse the data by separately solving the smaller-the-better problem of the lower limit and the larger-the-better problem of the upper limit. Are there any contradictions in the conclusions?

(c) Compare the two solutions with separate objective functions with the one in Example 6.4.

(d) The operating window turned out to be $[235.7°; 251.7°]$, but only factors D, H, and J were considered. Even if the other factors are statistically insignificant for the window width, they may be significant for the window position. Estimate the window width with this in mind.

(e) The window $[235.7°; 251.7°]$ was obtained using only the point estimates of the means and standard deviations. What will this window be if the uncertainties of these estimates are taken into account?

Table 6.9   The amount of white spots in Exercise 6.3.

	Control			Noise		Replicate			
	A	B	C	D	E	1	2	3	4
1	−1	−1	−1	−1	−1	7.5	12.3	8.0	9.3
2	1	−1	−1	−1	1	3.9	2.5	2.7	2.2
3	−1	1	−1	−1	1	11.8	13.1	14.8	11.6
4	1	1	−1	−1	−1	4.0	3.3	6.5	4.7
5	−1	−1	1	−1	1	14.3	6.6	11.2	8.2
6	1	−1	1	−1	−1	2.3	1.9	4.9	3.9
7	−1	1	1	−1	−1	4.0	6.6	7.0	5.0
8	1	1	1	−1	1	3.1	2.4	4.8	1.4
9	−1	−1	−1	1	1	18.7	24.9	22.2	17.0
10	1	−1	−1	1	−1	7.5	6.6	5.2	10.1
11	−1	1	−1	1	−1	13.6	14.3	15.4	15.9
12	1	1	−1	1	1	8.7	8.3	10.6	8.3
13	−1	−1	1	1	−1	13.0	13.7	7.9	14.6
14	1	−1	1	1	1	4.7	9.3	6.0	2.6
15	−1	1	1	1	1	8.9	10.9	13.7	13.6
16	1	1	1	1	−1	5.2	3.6	5.8	4.0
17	0	0	0	0	0	1.6	1.4	0.7	0.7

References

Peace GS (1993) *Taguchi methods*. Addison-Wesley.

Phadke MS (1989) *Quality engineering using robust design*. Prentice Hall.

Wu J and Hamada M (2000) *Experiments. Planning, analysis, and parameter design optimization*. John Wiley & Sons, Inc.

Wu Y and Wu A (2000) *Taguchi methods for robust design*. ASME Press, Fairfield, New Jersey.

7

Regression for robust design

7.1 Graphical techniques

Designed experiments are an extremely efficient way to collect data, especially for robust design activities. However, there are instances when a designed experiment is not an option, or at least when some variables cannot be fully controlled. To illustrate this, consider the data of Table 7.1 where the noise C is intended to take the values -1 and 1. However, even if this is intended, the experimenter does not succeed in accomplishing it. The situation is rather typical. Some variables are not fully controllable in the experiment. Since everything so far in the book has been based on results from factorial experiments, the methods that have been applied cannot fully be used, especially not the graphical techniques. However, the fundamental idea is still valid, namely to keep

$$\sigma_y^2 = \left(g'(z)\right)^2 \sigma_z^2$$

small by focusing on $g'(z)$ and to find levels of the control factors $x_1, \ldots, x_p$ that minimize $\left(g'(z)\right)$,

$$\min_{x_1, \ldots, x_p} \left(\frac{\partial g(x_1, \ldots x_p, z_1, \ldots, z_q)}{\partial z_i}\right)^2, \quad i = 1, \ldots, q.$$

One way to approach this problem is to build a regression model and then plot the control-by-noise interactions of the fitted model. This may seem easy, but there are several difficulties attached to it. First of all, if there are variables that cannot be fully controlled in the experiment, it must still be possible to observe them, otherwise a regression approach cannot be used. If they are not even observable, the only

Statistical Robust Design: An Industrial Perspective, First Edition. Magnus Arnér.
© 2014 John Wiley & Sons, Ltd. Published 2014 by John Wiley & Sons, Ltd.
Companion website: www.wiley.com/go/robust

Table 7.1   Regression data with noise and control.

Control factors		Noise	Response
A	B	C	y
−1	−1	−0.8	22.1
1	−1	0.4	22.7
−1	1	1.1	24.3
1	1	−1.3	8.7
−1	−1	−0.7	22.3
1	−1	−0.5	12.1
−1	1	0.1	23.1
1	1	1.7	26.3

remedy is to use dispersion models, as in Section 2.3, at least if the unobservable variable is noise. Another issue is the difficulties with model building activity such as collinearity, but this problem will not be discussed here since it does not differ from the difficulties always present in statistical model building and is not specific to robust design.

A third difficulty is in the graphical presentation of the results. If, say, the interaction between control factor x_A and noise variable x_C is plotted, the value of control factor x_B has to be fixed. It is not always obvious what this fixed value should be.

Example 7.1   For the data of Table 7.1, the model

$$y = 20.0 - 3.0x_A + 3.7x_C + 2.6x_A x_C + \varepsilon \qquad (7.1)$$

is obtained. It is worth noticing that there is one control-by-noise interaction in the model, the one including variables x_A and x_C. Note also that variable x_B is not in this model at all, so the last one of the issues mentioned above will not appear. The parameter x_B does not have to be fixed in an interaction plot since it does not have a statistically significant effect on the response y. For the values $x_A = -1$ and $x_A = 1$ of the control parameter x_A we obtain the equation

$$y = 23.0 + 1.1x_C + \varepsilon$$

if $x_A = -1$ and

$$y = 17.0 + 6.3x_C + \varepsilon$$

if $x_A = 1$. Figure 7.1 reveals that the control factor x_A should be set to $x_A = -1$ to obtain robustness.

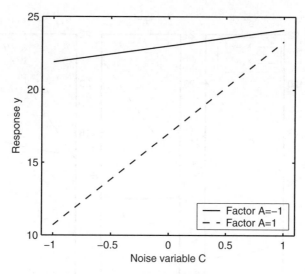

Figure 7.1   A factor effect plot for the fitted model. Data from Table 7.1.

Having come so far, a question may arise. Why bother about controlling the noise at all in the experimentation when it is possible to reach the conclusion (to set $x_A = -1$) so easily anyway? One answer is the number of runs. The data of Table 7.1 has actually been adjusted to make the analysis easy. In general, it cannot be expected that eight runs is enough if there are three variables. With regression variables, a larger number of runs are generally needed. Designed experiments are used since they provide more information than other experiments for a given number of runs. Another issue is that prior to the experiment it is unknown whether the noise just accidentally will encompass the entire range of interest. Thus, it will not be known beforehand if the experiment will generate the necessary information. However, this example proves that it is possible to learn about robustness even if regression variables are involved.

Example 7.2   Let us consider a slightly more complicated example. Assume that a regression model is fitted to data and the regression equation

$$y = 17.0 + 1.7x_A + 1.3x_B + 1.1x_C - 0.7x_{AB}$$
$$+1.1x_{AC} + 0.3x_{BC} + 0.1x_C^2 + \varepsilon \tag{7.2}$$

is obtained, where x_A and x_B are control parameters and x_C is noise. Apparently, there are some control-by-noise interactions. The way in which an interaction plot of A-by-C will look depends on the value taken by factor B. For $x_B = -1$, the relation is

$$y = 13.3 - 0.3x_C + 0.1x_C^2 + \varepsilon \quad \text{if } x_A = -1$$
$$y = 18.1 + 1.9x_C + 0.1x_C^2 + \varepsilon \quad \text{if } x_A = 1.$$

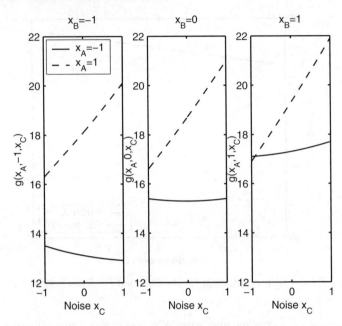

Figure 7.2   Interaction plots for the model of Equation (7.2).

For $x_B = 0$, on the other hand, the relation is

$$y = 15.3 + 0.1x_C^2 + \varepsilon \quad \text{if } x_A = -1$$
$$y = 18.7 + 2.2x_C + 0.1x_C^2 + \varepsilon \quad \text{if } x_A = 1.$$

This is sketched in Figure 7.2. In this case it is obvious that variable x_A should take the value $x_A = -1$ to become robust against the noise.

7.2   Analytical minimization of $\left(g'(z)\right)^2$

Example 7.3   The solution to a robust design problem with uncontrollable noise in the experiments does not always turn out as easily as in Examples 7.1 and 7.2. The remedy is then to go back to what actually is intended, namely to minimize $\left(g'(z)\right)^2$ in the expression

$$\sigma_y^2 = \left(g'(z)\right)^2 \sigma_z^2$$

where z is the noise. In Example 7.2, $z = x_C$ and we have

$$g'(z) = \frac{\partial g(x_A, x_B, z)}{\partial z} = 1.1 + 1.1x_A + 0.3x_B + 0.2z$$

and

$$\left(g'(z)\right)^2 = \left(1.1 + 1.1x_A + 0.3x_B + 0.2z\right)^2.$$

This equation takes its minimum when

$$\frac{\partial \left(g'(z)\right)^2}{\partial x_A} = 0$$

$$\frac{\partial \left(g'(z)\right)^2}{\partial x_B} = 0.$$

We obtain

$$\frac{\partial \left(g'_z\right)^2}{\partial x_A} = 1.1 \times 2 \left(1.1 + 1.1x_A + 0.3x_B + 0.2z\right) = 0$$

$$\frac{\partial \left(g'_z\right)^2}{\partial x_B} = 0.3 \times 2 \left(1.1 + 1.1x_A + 0.3x_B + 0.2z\right) = 0.$$

The solution to this equation depends on the value of x_C. A natural choice is to set this to its expected value. Let us assume that $\mu_z = 0$. In this case, the solution will not be a single value but a solution space, namely

$$x_A = \frac{-1.1 - 0.3x_B}{1.1}.$$

Thus, the solution is

$$x_A = -\frac{8}{11} \quad \text{if } x_B = -1$$

$$x_A = -1 \quad \text{if } x_B = 0$$

$$x_A = -\frac{14}{11} \quad \text{if } x_B = 1.$$

7.3   Regression and crossed arrays

It is not obvious that crossed arrays should be discussed at all in a chapter about regression. It is not even obvious what to model with regression. One application of regression for crossed arrays has already been studied in this book, one with the signal as the regressor. Besides this, regression may be used in several other ways:

- When terms in the outer array are not fully controlled in the experimentation, but the terms of the inner array are.

- When terms in the inner array are not fully controlled, but the terms of the outer array are.

- When neither the terms of the inner array nor the terms of the outer array are fully controlled.

Example 7.4   Consider an electrical window regulator of a car. An electrical motor drives the motion of the window elevator. Suppose that the electrical power UI is the signal and the speed v of the window is the response. There are weather strips in the door for sealing and to guide the window. Suppose that wear and dust intrusion in the weather strips make the window elevation require more power when the system gets older. Age will therefore be considered as noise.

In this example, only the voltage U of the signal but not the current I can be controlled in the experiment. The motor will use the current it requires. This means that the signal M cannot be fully controlled. Nevertheless, the slope β in the equation

$$y = \beta M + \varepsilon$$

can still be estimated since regression always works. There is an engineering way of viewing this as well. If the signal is something that the customer applies, like brake torque, then it cannot have any uncontrollable part because it would make the function unpredictable. However, if the signal is something that should be used to the smallest possible amount, like the power in this example, then it may sometimes contain uncontrollable elements.

For the electrical motor, assume that the true but unknown model is

$$y = E[Y|M] + \varepsilon = \beta M + \varepsilon$$
$$\beta = 0.10 + 0.01x_1 - 0.01x_3 - 0.02z_1$$
$$\log\left(\frac{\sigma}{E[Y|M]}\right) = -2 + 0.02x_1 + 0.02x_2 + 0.01z_1$$

and the result of Table 7.2 is obtained. This type of model is in line the ideas of Taguchi, giving higher standard deviations with higher values of the response.

The slope β and standard deviation σ, or some function of σ such as $\log(\sigma)$, can be estimated for each one of these four combinations of Table 7.2, as shown in Table 7.3 and Figure 7.3. Another way to quantify the variation is to have $\log(\sigma/\mu)$ as the response, in the same manner as the data was generated. However, it is a bit more complicated and will be discussed later on.

Let

$$X = \begin{pmatrix} 1 & -1 & -1 & -1 \\ 1 & 1 & -1 & 1 \\ 1 & -1 & 1 & 1 \\ 1 & 1 & 1 & -1 \end{pmatrix}$$

Table 7.2   Window elevator data of Example 7.4.

Control factors				Noise factor (age)			
x_1	x_2	x_3		-1	1	-1	1
			Signal	42	51	81	90
-1	-1	-1	Response	3.5	4.3	9.9	6.6
			Signal	37	41	72	74
1	-1	1	Response	4.1	2.8	8.6	6.2
			Signal	40	51	77	85
-1	1	1	Response	3.8	2.9	7.7	5.0
			Signal	33	39	70	72
1	1	-1	Response	4.0	4.6	10.4	8.5

denote the design matrix. The estimated effects of the control factors on $\log(\sigma)$ are

$$\hat{\theta} = (X'X)^{-1}X' \begin{pmatrix} 0.54 \\ 0.26 \\ 0.43 \\ -0.05 \end{pmatrix} = \begin{pmatrix} 0.295 \\ -0.191 \\ -0.105 \\ 0.052 \end{pmatrix}$$

and on the slope

$$\hat{\phi} = (X'X)^{-1}X' \begin{pmatrix} 0.093 \\ 0.098 \\ 0.076 \\ 0.130 \end{pmatrix} = \begin{pmatrix} 0.099 \\ 0.015 \\ 0.004 \\ -0.012 \end{pmatrix}.$$

Assume a two-step approach of first minimizing variation and then maximizing the efficiency. Term selection is made using forward selection with an F-to-enter test. As

Table 7.3   Estimated parameters for the window elevator.

Control factors			Estimates			
x_1	x_2	x_3	$\hat{\beta}$	s	$\log s$	$\log\left(s/\hat{\beta}\right)$
-1	-1	-1	0.093	1.71	0.54	2.92
1	-1	1	0.098	1.29	0.26	2.28
-1	1	1	0.076	1.54	0.43	3.02
1	1	-1	0.130	0.95	-0.05	1.99

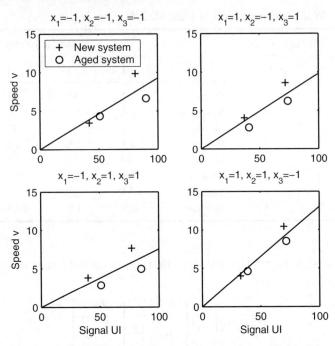

Figure 7.3   Fitted regression lines for the data of Table 7.2.

a measure of variation, consider $\log(\sigma)$. Then, both the model for the variation and for the average will be empty,

$$\log(\sigma) = 0.30 + v$$
$$\beta = 0.099 + \varepsilon$$

(the p values are 0.15 and 0.24 respectively for entering one variable into the model, x_1 in both cases).

Thus, with the data analysis performed in this way, it cannot be proved that any one of the control factors can be used for the purpose of getting an efficient and robust window regulator. One reason is of course the degrees of freedom. With only four data points it is always a difficulty to obtain a good model.

However, the data of the example were generated using another model, namely

$$\log\left(\frac{\sigma}{E[Y|M]}\right) = -2 + 0.02x_1 + 0.02x_2 + 0.01z_1.$$

Since this model has been used to generate the data, it is natural to analyse the data along the same line. It turns out to be slightly more complicated, since one response has to be calculated for each one of the 16 runs. The design matrix is

$$
X = \begin{pmatrix}
1 & -1 & -1 & -1 \\
1 & -1 & -1 & -1 \\
1 & -1 & -1 & -1 \\
1 & -1 & -1 & -1 \\
1 & 1 & -1 & 1 \\
1 & 1 & -1 & 1 \\
1 & 1 & -1 & 1 \\
1 & 1 & -1 & 1 \\
1 & -1 & 1 & 1 \\
1 & -1 & 1 & 1 \\
1 & -1 & 1 & 1 \\
1 & -1 & 1 & 1 \\
1 & 1 & 1 & -1 \\
1 & 1 & 1 & -1 \\
1 & 1 & 1 & -1 \\
1 & 1 & 1 & -1
\end{pmatrix}
$$

so that each row of the original inner matrix is repeated four times. The response is $y_{ij} = \log(s_i/(\hat{\beta}_i M_{ij}))$, where index i denotes the row of the original inner array and j the column of the outer array. In order to carry out the analysis, the data are stacked into a common column,

$$
Y = \begin{pmatrix}
\log\left(\dfrac{1.71}{0.093 \times 42}\right) \\
\vdots \\
\log\left(\dfrac{0.95}{0.0130 \times 72}\right)
\end{pmatrix} = \begin{pmatrix}
-0.82 \\
\vdots \\
-2.29
\end{pmatrix}.
$$

The estimated coefficients become

$$
(X'X)^{-1}(X'Y) = \begin{pmatrix}
-1.41 \\
-0.06 \\
-0.31 \\
-0.03
\end{pmatrix}.
$$

It is obvious that these estimates differ considerably from the values in the simulation model. This is not just a coincidence or bad luck. The generated and estimated standard deviations are not the same, as mentioned in Chapter 5. The so-called standard deviation in crossed arrays (which is the one estimated here) is a sum of the noise effect and the variation not encountered by the noise variables,

$$\sigma_{CA}^2 = \sigma^2 + g((z - \bar{z})^2)$$

where the index CA denotes crossed arrays. In the simplest case, it is

$$\sigma_{CA}^2 = \sigma^2 + K \sum (z - \bar{z})^2.$$

Note 7.1   Another way to handle the current in this case is to consider it as a part of the response, so that the signal just is the voltage U but the response is v/I ($v =$ speed). The relation between the signal and response would thus look not like

$$v = \beta UI + \varepsilon$$

but like

$$\frac{v}{I} = \beta U + \varepsilon.$$

One benefit would be conceptual; it may be easier to grasp the problem if the signal does not contain an element of randomness.

Note 7.2   It is fairly common that the signal consists of several parts, as the electrical power UI in this case. The response may depend on how this signal is constituted, that is it may matter if

$$UI = 2V \times 1A \text{ or}$$
$$UI = 1V \times 2A.$$

This distribution between the different elements of the signal may then be treated as a control factor if the elements of the signal are controllable or a noise factor when the current cannot be controlled (as here). For simplicity, it may be easier to consider the current value rather than the ratio as a factor, so that the P diagram of Figure 7.4 is obtained.

A signal that is not fully controllable is only occasionally a problem. If correctly analysed, useful and correct conclusions will mostly be drawn. It may of course happen that the signal will span completely different intervals for different rows of the inner array and that mistakes are made for this reason. However, the problem is limited and rare. The ease to deal with a noncontrollable signal is, by the way, one reason to sometimes use crossed arrays rather than combined arrays.

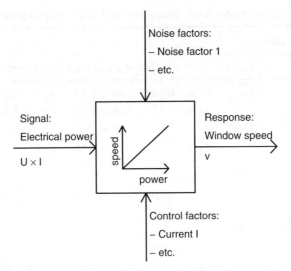

Figure 7.4   Sketch of a P diagram for a window regulator.

Assume that it is the other type of term in the outer array, the noise, that cannot be fully controlled. If it cannot be controlled or observed at all but the value it takes is a random number from its probability distribution, we are back to replicates as in traditional DOE. Unless there is a signal, it does then not make sense to consider it as an inner and outer array at all. Even if the noise can partially be controlled, the problem differs so little from the dispersion problem of Section 2.3 that a deeper discussion is unnecessary. However, if it is known in advance that the noise variables will be observable but not controllable in the test, then it is better to plan the experiment in terms of a combined array and use the techniques of Sections 7.1 and 7.2. If it is discovered during or after a crossed array experiment that the noise could not be controlled (but observed), then it is usually better to rewrite it into a combined array, even though this experimental array is far from optimal.

It is when the uncontrollable part is in the inner array that traditional regression techniques become useful from a theoretical point of view, but this is an uncommon situation in practice.

7.3.1   Regression terms in the inner array

Even if it is theoretically possible that the factors in the outer array, the noise, can be fully controlled in the experiment but not the factors of the inner array, it is a highly unusual situation. It is more likely that none of them can be fully controlled, so that data as in Table 7.4 are obtained. The natural way to approach the problem is to rewrite the data in 16 rows and consider it as regression data from a combined array.

Table 7.4   The actual factor levels in an inner and outer array experiment.

Control variables						
x_1	x_2	x_3	Noise variables z and responses y			
-1.14	-0.87	-0.58	$z_1 = -0.80$ $z_2 = -1.12$ $y = 14.1$	$z_1 = 1.17$ $z_2 = -1.51$ $y = 14.9$	$z_1 = -0.87$ $z_2 = 1.12$ $y = 15.3$	$z_1 = 1.17$ $z_2 = 1.01$ $y = 16.0$
0.98	-1.31	0.75	$z_1 = -1.23$ $z_2 = -0.89$ $y = 12.0$	$z_1 = 0.83$ $z_2 = -1.53$ $y = 13.9$	$z_1 = -0.99$ $z_2 = 1.14$ $y = 13.7$	$z_1 = 0.98$ $z_2 = 0.21$ $y = 15.9$
-0.99	1.01	1.09	$z_1 = -0.10$ $z_2 = -1.12$ $y = 17.1$	$z_1 = 0.88$ $z_2 = 1.27$ $y = 17.7$	$z_1 = -1.09$ $z_2 = 1.17$ $y = 17.6$	$z_1 = 1.12$ $z_2 = 1.10$ $y = 18.2$
0.72	0.57	-1.03	$z_1 = -0.87$ $z_2 = -0.87$ $y = 15.7$	$z_1 = 0.97$ $z_2 = 1.31$ $y = 16.9$	$z_1 = -0.37$ $z_2 = 1.03$ $y = 17.3$	$z_1 = 0.80$ $z_2 = 1.18$ $y = 19.6$

Exercises

7.1 Build a model from the data of Table 7.1 and compare it to Equation (7.1).

7.2 In a production process, some process parameters are difficult to control exactly. In parts of the chemical industry it can typically be pressure and temperature values. The nominal value can be used as the control, the deviance from the nominal noise. Since this deviance is small, it is really difficult to control it to perfection, even in a laboratory. The obtained pressure in the test has a tendency to differ somewhat from the intended, but it is possible to measure.

For an experiment in the process industry with the factors and factor levels of Table 7.5, the result of Table 7.6 was obtained. Analyse the data and explore how robustness against pressure disturbances can be obtained.

Table 7.5   Intended levels in the factorial design (Exercise 7.2).

Factor	Factor type	Low level	High level
A: Unspecified	Control	-1	1
B: Nominal pressure	Control	1	2
C: Deviation from nominal pressure	Noise	-0.1	0.1

Table 7.6   Intended and actual factor levels in Exercise 7.2.

| Run | Intended experiment | | | Intended pressure | Obtained pressure | Response |
| | Control | | Noise | | | |
	A	B	C			
1	−1	1	−0.1	0.9	0.97	4.50
2	−1	1	0.1	1.1	1.17	4.89
3	−1	2	−0.1	1.9	1.89	6.37
4	−1	2	0.1	2.1	2.02	6.51
5	1	1	−0.1	0.9	0.91	4.90
6	1	1	0.1	1.1	1.07	5.89
7	1	2	−0.1	1.9	1.98	7.47
8	1	2	0.1	2.1	2.15	7.62

7.3 (a) Consider Example 7.4. Keeping in mind the fact that the slope should be maximized, could this type of problem be considered as larger-the-better?

   (b) Analyse the data expressing the variation in terms of $\log(s/\hat{\beta})$. Compare also with the discussion about larger-the-better in (a).

   (c) Suppose that the voltage was 5 V and 10 V respectively for the two first and two last runs on each row. Consider the current consumption as a part of the response and analyse the data in this way.

8

Mathematics of robust design

The purpose of this chapter is to provide the reader with some mathematical background to robust design, primarily to make robust design based on computer experiments easier. Some mathematical concepts of robust design are discussed, but the treatment will not be of any depth. The mathematics is just be touched upon in order to understand better other topics of robust design.

The chapter is divided into three parts. First, a notational system is defined. To some extent, it is the notations that have been used earlier on in this book. Then a number of optimality criteria for robustness are discussed. Finally, the variance is decomposed into its components by using ANOVA, and it will be seen how this can be beneficial for robust design.

8.1 Notational system

The response y is given by the equation

$$y = g(\mathbf{x}, \mathbf{z}, \mathbf{M}) + \varepsilon \tag{8.1}$$

where

$$\mathbf{x} = (x_1, \dots, x_p) \quad \text{are control variables, } p \geq 1$$
$$\mathbf{z} = (z_1, \dots, z_q) \quad \text{are noise variables, } q \geq 1$$
$$\mathbf{M} = (M_1, \dots, M_r) \quad \text{are signal variables, } r \geq 0.$$

The term ε may be omitted if all noise is modelled and thus included in the noise variables z. This is typically the case in virtual experimentation. Otherwise, the term ε will be assumed to be $N(0, \sigma)$ unless anything else is stated. The number of signals,

Statistical Robust Design: An Industrial Perspective, First Edition. Magnus Arnér.
© 2014 John Wiley & Sons, Ltd. Published 2014 by John Wiley & Sons, Ltd.
Companion website: www.wiley.com/go/robust

r, will take a value in the set $\{0, 1\}$ in this chapter. However, Taguchi often makes use of double signals and $r \in \{0, 1\}$ is not a general requirement. Occasionally, we will write

$$y = g(\mathbf{x}) + \varepsilon$$

with the same meaning as Equation (8.1), so the vector $\mathbf{x}$ will sometimes just mean the control variables and sometimes all variables. The context will give the meaning. The noise, z, is a random variable with the probability density function $f_Z(\mathbf{z})$.

A matrix representation rather than a functional one is often useful. Let us write

$$X = \begin{pmatrix} 1 & \vdots & X_x & \vdots & X_z & \vdots & X_M & \vdots & X_{xz} & \vdots & X_{xM} & \vdots & X_{zM} & \vdots & X_{xzM} \end{pmatrix}$$

where $\mathbf{1}$ is an $n \times 1$ vector whose all elements are 1. The subscript indicates the type of variable that is considered, so that

$$X_x = \begin{pmatrix} x_{11} & \cdots & x_{1p} \\ \vdots & \ddots & \vdots \\ x_{n1} & \cdots & x_{np} \end{pmatrix}$$

is the part of the design matrix with the control variables. This matrix may include polynomial terms and interactions between control parameters. The matrix X_{xz} is the matrix of control-by-noise interaction terms. The content of the rest of the elements of the design matrix X should be obvious. Using this notation, we can write

$$y = X\beta + \varepsilon.$$

8.2 The objective function

In Chapter 7 we made use of the partial derivatives with respect to the noise

$$u_i = g_i'(x, z) = \frac{\partial g(x, z)}{\partial z_i}$$

and concluded that robust design actually is a minimization problem,

$$\min_{x_1, \ldots, x_p} u_i^2, \quad i = 1, \ldots, q.$$

We will look at some alternatives to this and discuss some drawbacks and advantages with different criteria for optimal robustness.

Example 8.1 Consider the equation

$$y = g(x, z) = 2x - 1.7z + 0.3z^2 + 1.7xz - 0.2x^2z^2 \tag{8.2}$$

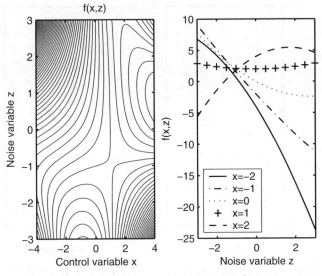

Figure 8.1   The function $f(x, z) = 2x - 1.7z + 0.3z^2 + 1.7xz - 0.2x^2z^2$.

which is illustrated in Figure 8.1 and where

$$x = \text{a control variable and}$$
$$z = \text{a noise variable.}$$

The noise z is a random variable,

$$Z \text{ is } N(\mu_z, \sigma_z), \quad \mu_z = 0, \quad \sigma_z = 1.$$

The objective in robust design is to minimize the sensitivity to noise. In principle, the problem of robustness is to find a value of x so that

$$u = \frac{\partial g}{\partial z} = -1.7 + 0.6z + 1.7x - 0.4x^2z$$

is minimized, or rather as close to zero as possible (Figure 8.2).

The value of x minimizing u will be a function of z which complicates the problem. In Chapter 7 this issue was approached by taking the derivative around the mean of z,

$$u(x, z)|_{z=\mu_z} = -1.7 + 0.6\mu_z + 1.7x - 0.4x^2\mu_z = -1.7 + 1.7x.$$

Solving the equation gives

$$u(x, z)|_{z=\mu_z} = 0 \Rightarrow x = 1.$$

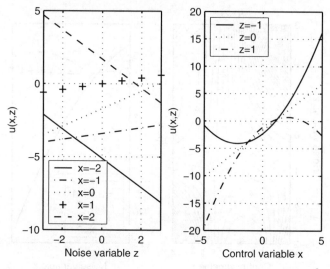

Figure 8.2   The function $u(x, z) = -1.7 + 0.6z + 1.7x - 0.4x^2z.$

However, there are other solutions to the robustness problem that could be more relevant. One is to consider the average sensitivity,

$$E[u(x, Z)] = \int_{-\infty}^{\infty} u(x, z) f(z) \mathrm{d}z$$

and solve the equation

$$E[u(x, Z)] = 0.$$

For Equation (8.2) we obtain

$$\begin{aligned} E[u(x, Z)] &= \int_{-\infty}^{\infty} \left(-1.7 + 0.6z + 1.7x - 0.4x^2 z \right) f(z) \mathrm{d}z \\ &= -1.7 + 1.7x = 0, \end{aligned}$$

so that $x = 1$ minimizes the sensitivity to noise. Yet another reasonable approach is to minimize

$$\begin{aligned} E[u^2] &= \int_{-\infty}^{\infty} \left(-1.7 + 0.6z + 1.7x - 0.4x^2 z \right)^2 f(z) \mathrm{d}z \\ &= (-1.7 + 1.7x)^2 + (0.6 - 0.4x^2)^2. \end{aligned} \tag{8.3}$$

It is minimized when

$$x \approx 1.05.$$

These three versions of optimality will be called optimal robustness types II to IV (type I will be defined later).

Definition 8.1   *A solution x is of optimal robustness type II if*

$$u(x, z)|_{z=\mu_z} = 0.$$

A solution x is of optimal robustness type III if

$$E[u(x, Z)] = 0.$$

A solution x is of optimal robustness type IV if

$$E\left[(u(x, Z))^2\right]$$

is minimized.

A drawback with types II and III is that they do not always exist. Their existence requires that the derivative takes the value zero for some value of x. In practice, this drawback is important since it is very common that there is a monotone relation between the noise z and the response y, no matter what the value of the control x is. We have even seen (Chapter 3) that the very idea of compounding noise factors is based on this monotony. However, if the response y is monotone in z, the derivative u will never be zero. Type IV does not have this drawback and is thus useful in more general situations.

It could be argued that all these optimality criteria are nonsense since the pdf $f_Z(z)$ hardly ever is known in real life. There might be a rough idea the distribution of z, but even this rough idea is often missing. Despite that, these optimality criteria are of some value, primarily from a conceptual point of view, but also in practice. Especially in cases when the experimentation is virtual and it is necessary to run an optimization algorithm.

Type II optimality has yet another drawback. As the following example shows, it may give very misleading results if the noise is a random variation around the nominal, as it is on so many occasions in manufacturing applications.

Example 8.2   Consider the function

$$g(x, z) = \begin{cases} -(x + z - 0.9)^2 & \text{if } 0 \le x + z \le 0.9 \\ -20(x + z - 0.9)^2 & \text{if } 0.9 < x + z \le 1 \end{cases}$$

where $\mu_z = 0$ (Figure 8.3). The function resembles the one with adhesion in Exercise 1.1 and is a typical example of how it might look if the noise is a random variation around the nominal. It is obvious that type II optimal robustness is achieved for $x = 0.9$, but it is also obvious that this solution is not at all robust. The conclusion is that type II optimality cannot be used for problems of this kind.

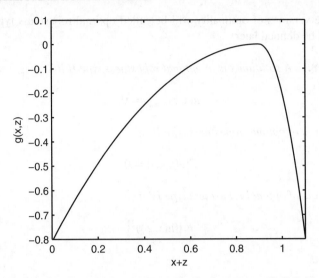

Figure 8.3   A situation for which type II optimal robustness will fail. The noise is a random variation around the nominal.

8.2.1   Multidimensional problems

In both the Examples 8.1 and 8.2, there was only one noise variable. When there are more noise variables, the problem becomes more difficult. Let us take an example.

Example 8.3   Consider the function

$$g\left(x, z_1, z_2\right) = -1.7z_1 + 1.7xz_1 + 0.8z_2 - 1.3xz_2.$$

If this is differentiated with respect to the noise we obtain

$$u_1 = \frac{\partial g}{\partial z_1} = -1.7 + 1.7x$$

$$u_2 = \frac{\partial g}{\partial z_2} = 0.8 - 1.3x$$

Suppose that the mean $\mu_{z_i} = 0$ for $i = 1$ and 2. Solving for type II optimal robustness, we obtain

$$u_1|_{z_1=\mu_{z_1}} = 0 \Rightarrow x = 1$$

$$u_2|_{z_2=\mu_{z_2}} = 0 \Rightarrow x = \frac{0.8}{1.3} \approx 0.61 \neq 1.$$

Obviously, the problem must be addressed in some other manner. The solution is not to be found directly in any one of the three optimality criteria of Definition 8.1 since they are all expressed in univariate terms.

A natural approach is to minimize the variance of g. This could be considered as an optimality criterion.

Definition 8.2 *A solution x is of optimal robustness type V if*

$$Var\left[g(x, Z)\right]$$

is minimized.

Let us apply this to our Example 8.3. Assume that

$$(Z_1, Z_2)' \text{ is } N\left(\begin{pmatrix} 0 \\ 0 \end{pmatrix}, \begin{pmatrix} \sigma_1^2 & 0 \\ 0 & \sigma_2^2 \end{pmatrix}\right).$$

Define

$$A = \begin{pmatrix} -1.7 + 1.7x \\ 0.8 - 1.3x \end{pmatrix}$$

$$\Sigma = \begin{pmatrix} \sigma_1^2 & 0 \\ 0 & \sigma_2^2 \end{pmatrix}$$

$$Z = \begin{pmatrix} Z_1 \\ Z_2 \end{pmatrix}.$$

We obtain

$$Var\left[g(x, Z_1, Z_2)\right] = Var\, A'Z = A'\Sigma A.$$

This is minimized when

$$\frac{\partial A'\Sigma A}{\partial x} = 0.$$

Some calculations give

$$A'\Sigma A = (-1.7 + 1.7x)^2\, \sigma_1^2 + (0.8 - 1.3x)^2\, \sigma_2^2$$

$$\frac{\partial A'\Sigma A}{\partial x} = 2 \times 1.7\, (-1.7 + 1.7x)\, \sigma_1^2 - 2 \times 1.3\, (0.8 - 1.3x)\, \sigma_2^2.$$

If $\sigma_1^2 = \sigma_2^2 = 1$, this equals zero when

$$x = \frac{2 \times 1.7^2 + 0.8 \times 2.6}{2 \times 1.7^2 + 2.6 \times 1.3} = \frac{7.86}{9.16} \approx 0.86.$$

8.2.2   Optimization in the presence of a signal

If there are three types of variables, control, noise, and signal, a two-step approach is appropriate to use. First minimize the sensitivity to noise,

$$\min_x u_z = \frac{\partial g(x, z, M)}{\partial z}$$

and then use the control parameters x not fixed to minimize sensitivity to noise in order to maximize the sensitivity to the signal,

$$\max_x u_M = \frac{\partial g(x, z, M)}{\partial M}.$$

Example 8.4   Consider the function

$$y = 10M + 1.7z + 1.7x_1z + 3x_1M - 0.7x_2M,$$
$$-2 \le x_i \le 2, \qquad i \in \{1, 2\}.$$

The sensitivity to noise is minimized when

$$u_z = \frac{\partial y}{\partial z} = 1.7 + 1.7x_1 = 0,$$

that is when

$$x_1 = -1.$$

The next step is to maximize the sensitivity to the signal,

$$\max u_M = \frac{\partial y}{\partial M} = 10 + 3x_1 - 0.7x_2.$$

Since x_1 is already fixed to the value -1, we only have the control factor x_2 to play with. The sensitivity u_M is maximized when

$$x_2 = -2$$

when it takes the value

$$u_M = 10 - 3 + 1.4 = 8.4.$$

8.2.3   Matrix formulation

A robust design problem can be written in matrix form as

$$
\begin{aligned}
y &= \beta_0 + X_x\beta_x + X_z\beta_z + X_{xz}\beta_{xz} + \varepsilon \\
&= \beta_0 + X_x\beta_x + X_z\beta_z + X_x B X_z' + \varepsilon.
\end{aligned}
\tag{8.4}
$$

(see, for example, Myers *et al.*, 1992). Given the levels of the control factor x, we obtain the variance

$$
\begin{aligned}
Var(Y) &= Var\left(X_z\beta_z + X_x^* B X_z\right) + \sigma_\varepsilon^2 \\
&= \left(\beta_z + X_x^* B\right) \Sigma \left(\beta_z + X_x^* B\right)' + \sigma_\varepsilon^2
\end{aligned}
$$

where X_x^* is a fixed set of control factor levels, X_z the noise random variables, and

$$
\Sigma = Cov(X_z).
$$

The robust optimization represents the minimization of

$$
\left(\beta_z' + X_x^* B\right) \Sigma \left(\beta_z' + X_x^* B\right)'.
\tag{8.5}
$$

Written in this form, Equation (8.2) becomes

$$
y = \beta_0 + X_x\beta_x + X_z\beta_z + X_x B X_z' + \varepsilon
$$

where

$$
X_x = \begin{pmatrix} x & x^2 \end{pmatrix}, \quad X_z = \begin{pmatrix} z & z^2 \end{pmatrix},
$$

and

$$
\beta_0 = 0, \quad \beta_x = \begin{pmatrix} 2 \\ 0 \end{pmatrix}, \quad \beta_z = \begin{pmatrix} -1.7 \\ 0.3 \end{pmatrix}, \quad B = \begin{pmatrix} 1.7 & 0.0 \\ 0.0 & -0.2 \end{pmatrix}.
$$

Since

$$
\Sigma = \begin{pmatrix} Var(Z) & Cov(Z, Z^2) \\ Cov(Z^2, Z) & Var(Z^2) \end{pmatrix} = \begin{pmatrix} \sigma_z & 0 \\ 0 & 2\sigma_z^4 \end{pmatrix} = \begin{pmatrix} 1 & 0 \\ 0 & 2 \end{pmatrix}
$$

the optimization problem is to find the value of x minimizing

$$
\begin{aligned}
&\left(\begin{pmatrix} -1.7 & 0.3 \end{pmatrix} + \begin{pmatrix} x & x^2 \end{pmatrix} \begin{pmatrix} 1.7 & 0.0 \\ 0.0 & -0.2 \end{pmatrix} \right) \Sigma \left(\begin{pmatrix} -1.7 & 0.3 \end{pmatrix} + \begin{pmatrix} x & x^2 \end{pmatrix} \begin{pmatrix} 1.7 & 0.0 \\ 0.0 & -0.2 \end{pmatrix} \right)' \\
&= (-1.7 + 1.7x)^2 + 2(0.3 - 0.2x^2)^2.
\end{aligned}
\tag{8.6}
$$

This resembles Equation (8.3) but is not identical. Equation (8.6) attains its minimum for $x \approx 1.025$ and Equation (8.3) for $x \approx 1.05$. In general, we cannot assume that the expected value of the squared derivative will be identical to the variance of the original expression. Thus, in this case

$$E\left[\left(\frac{\partial g(x, Z)}{\partial Z}\right)^2\right] \neq E\left[(g(x, Z) - E[g(x, Z)])^2\right]. \tag{8.7}$$

This difference can be explained using Taylor expansions, and it turns out that optimal robustness type IV is simply a first order approximation. The value of x that minimizes Equation (8.5) is thus the optimal solution that may be considered as the most natural one.

Definition 8.3 *A solution x is of optimal robustness type I if*

$$\left(\beta'_z + X_x B\right) \Sigma \left(\beta'_z + X_x B\right)'$$

is minimized.

If there are dispersion effects, the matrix model can be expressed as

$$y = \beta_0 + X_x \beta_x + X_z \beta_z + X_x B X'_z + \varepsilon, \quad \varepsilon \sim N(0, \sigma(x, z))$$
$$\log(\sigma) = \theta_0 + X_x \theta_x + X_z \theta_z.$$

The dispersion model can, of course, occasionally include control-by-noise interaction terms as well.

Which representation to use, the matrix representation or the general function, depends on the application and the way of data collection. If the data are collected through physical experimentation, then the matrix representation is often more appropriate. The experimental result is used to estimate the unknown coefficients in the matrix model. On the other hand, if the experiments are virtual, then the functional representation may be more appropriate.

In Example 8.4, the optimization turned out to be easy since there is one control variable that interacts with the signal but not with the noise, namely x_2. This makes it possible to perform the two-step approach of first minimizing sensitivity to noise, and then maximizing sensitivity to the signal. Such luck is not unusual since most variables are involved in very few interactions. A variable like x_1, which has substantial interaction with both the noise and the signal, is the less common type of variable, not x_2. However, it happens now and then that no variables are left over once the minimization of the sensitivity to noise is accomplished. The two-step approach will then fail and both steps must somehow be taken simultaneously. One possibility is then to use an alternative optimality concept, Pareto optimality.

8.2.4   Pareto optimality

Vilfredo Pareto was an Italian economist in the late 19th and early 20th century. Pareto optimality, which is named after him, is a situation when available resources cannot be reallocated to make it better for someone without making it worse for someone else. In recent years this concept has gained interest in engineering, where it is common to have several objective functions to optimize simultaneously, but where the trade-off between them is far from obvious. The concept is widely used in computer experiments, and the application is not limited to robust design.

Example 8.5   In this example, we will apply Pareto optimality to robust design. Consider the two functions

$$y_1 = 10 + 1.7z + 1.7x_1z - 0.3x_2z, \qquad -1 \leq x_1, x_2 \leq 1$$
$$y_2 = 3 + 0.8z - 1.3x_1z + 0.9x_2z, \qquad -1 \leq x_1, x_2 \leq 1.$$

We will calculate the Pareto optimal set in the sense of type IV optimal robustness for these functions. Recall that type IV optimal robustness is obtained when

$$E\left[(u(x, Z))^2\right]$$

is minimized. In this case, u will be independent of z, and

$$u_1^2 = \left(1.7 + 1.7x_1 - 0.3x_2\right)^2$$
$$u_2^2 = \left(0.8 - 1.3x_1 + 0.9x_2\right)^2. \tag{8.8}$$

In Figure 8.4, the two functions of Equation (8.8) are sketched. A grid of points is calculated and the Pareto optimal points of this grid are marked. Using a tighter grid will show that the Pareto optimal set is a connected area.

We will not go into the details on how to calculate a Pareto optimal set. In this specific example it has been done in a simple way: the values of the two functions have been calculated for a grid of points, and the values are just compared to each other. Putting this into action will of course require some kind of algorithm, but it is outside the scope of this book. Pareto optimality is of specific interest in design and analysis of computer experiments, specifically robust design experiments, and useful algorithms are implemented in all major software packages for this purpose.

Consider Figure 8.5. The Pareto optimal points are marked. They are located along a front called the Pareto front. What we see is the values of $u_1^2(x_1, x_2)$ and $u_2^2(x_1, x_2)$. Even if two such points are close to each other, they may be far from each other when it comes to how the design looks like. From a design point of view, this observation is of primary interest. It means that we can point out designs that are fairly different from each other with regards to how they are designed, but similar in terms of the response we get from it. It gives us a possibility to choose.

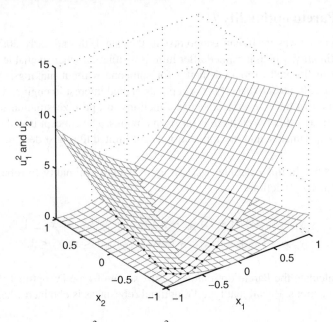

Figure 8.4   The functions $u_1^2(x_1, x_2)$ and $u_2^2(x_1, x_2)$. The set of Pareto optimal solutions in the grid of points is marked.

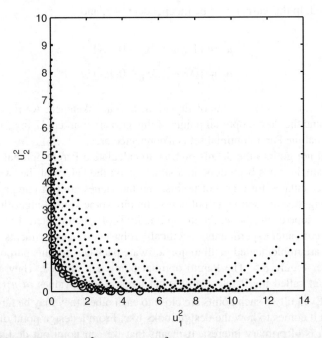

Figure 8.5   The values of $u_1^2(x_1, x_2)$ and $u_2^2(x_1, x_2)$ for the grid of calculated points. The points on the Pareto front are specifically marked.

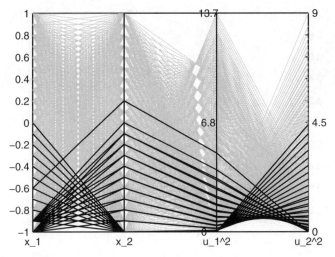

Figure 8.6   Parallel coordinate chart for Example 8.5. The Pareto optimal set is marked with bold lines.

A graphical tool that can be used to simplify the choice is the so-called parallel coordinate chart. Such a chart is sketched in Figure 8.6.

Example 8.6   Using parallel coordinate chart makes it easy to include restrictions, both on the response and the parameters. Suppose that Example 8.5 is adjusted a little, so that the problem is to find the Pareto optimal set minimizing

$$u_1^2 = \left(1.7 + 1.7x_1 - 0.3x_2\right)^2$$
$$u_2^2 = \left(0.8 - 1.3x_1 + 0.9x_2\right)^2$$

subject to

$$u_1^2 \leq 5$$
$$u_2^2 \leq 1.$$

Then, only 17 of the 31 designs of the Pareto optimal set remain, as illustrated in Figure 8.7. Some of these solutions are a fair distance away from each other from a design point of view, for example

$$\left(x_1, x_2\right) = (-1, -0.8)$$

and

$$\left(x_1, x_2\right) = (-0.6, 0.2).$$

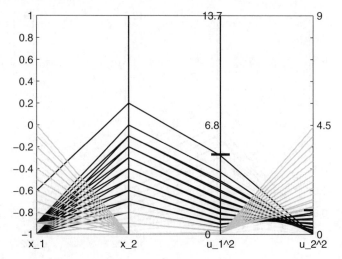

Figure 8.7   Parallel coordinate chart for Example 8.6. Only the Pareto optimal set is sketched. Those fulfilling the subjective functions are in solid lines.

It is thus possible to present a number of design solutions to the project leaders. If the solutions are fairly remote from each other from a design point of view but close from a robustness point of view, it will provide the project leaders with a possibility to choose.

8.3   ANOVA for robust design

Analysis of variance, ANOVA, can sometimes reveal an enormous amount of beneficial information in how to design to obtain robustness. It has a strong tradition in the Taguchi School and is thoroughly covered in the book by Wu and Wu (2000). The perspective here will, however, be slightly different from that of Wu and Wu (2000). An important topic will be functional ANOVA, a concept developed primarily by Sobol during the last few decades. Functional ANOVA has become an important issue in the analysis of data from computer experiments, and we will use it for this purpose in Chapter 9. However, before rushing into functional ANOVA, let us take a look at traditional ANOVA.

8.3.1   Traditional ANOVA

Let us for a moment leave robust design and just consider factors as factors in general. Assume that we have the data of Table 8.1.

The common way to analyse this set of data using ANOVA would then be to set up the model

$$y_{ij} = \beta_0 + \beta_1 x_i + \varepsilon_{ij}, \quad \varepsilon \text{ is } N(\mu, \sigma), \quad i = 1, \ldots, m, \quad j = 1, \ldots, n$$

Table 8.1 Data for an ANOVA analysis.

x	1	1	1	1	2	2	2	2
y	3.57	2.33	4.13	4.29	5.85	8.19	8.19	6.69

and divide the variation in y into the sources of variation, namely the regressor x and the random variation σ. For the data of Table 8.1, $m = 2$ and $n = 4$. We will divide the sum of squares,

$$SS_{TOT} = \sum_{i=1}^{m} \sum_{j=1}^{n} \left(y_{ij} - \bar{y}_{..} \right)^2$$

into its components. These components are SS_x and SS_{RES},

$$SS_{TOT} = SS_x + SS_{RES}$$

where

$$SS_x = n \sum_{i=1}^{m} \left(\bar{y}_{i.} - \bar{y}_{..} \right)^2 \text{ and } SS_{RES} = \sum_{i=1}^{m} \sum_{j=1}^{n} \left(y_{ij} - \bar{y}_{i.} \right)^2 .$$

For the data of Table 8.1, we obtain

$$SS_{TOT} = \sum_{i=1}^{m} \sum_{j=1}^{n} \left(y_{ij} - \bar{y}_{..} \right)^2 = 33.82$$

$$SS_x = n \sum_{i=1}^{m} \left(\bar{y}_{i.} - \bar{y}_{..} \right)^2 = 27.68$$

$$SS_{RES} = \sum_{i=1}^{m} \sum_{j=1}^{n} \left(y_{ij} - \bar{y}_{i.} \right)^2 = 6.14.$$

This result can be represented either graphically (Figure 8.8), or in terms of an ANOVA table (Table 8.2).

An example of ANOVA as simple as this will of course not help us in making a product robust against noise. With only one factor in the analysis, there is not even room for both control and noise factors. However, it will help us understand functional ANOVA in a better way since it will give us something to compare with; we will see that each and every term in functional ANOVA has a corresponding term in traditional ANOVA.

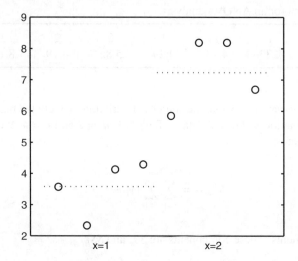

Figure 8.8   The data of Table 8.1 with the group averages $\bar{y}_1 = 3.58$ and $\bar{y}_2 = 7.30$ marked.

8.3.2   Functional ANOVA

In contrast to traditional ANOVA, functional ANOVA does not deal with data from an experiment, at least not directly. It deals with a mathematical function. The parameters and coefficients of this function can be estimated from experimental data. However, just like traditional ANOVA, it decomposes the total variation into the contribution from each source. It may seem as if this difference is subtle, and it might very well be, but there is one difference of importance. In traditional ANOVA the contribution from a single variable x is divided into its linear, quadratic, cubic contribution, and so on. This is not the case in functional ANOVA. All of these polynomial contributions are treated in one term.

Example 8.7   Consider the function

$$g(x_1, x_2) = 2x_1^2 + x_1 x_2, \qquad 0 \le x_i \le 1, \qquad i = 1, 2.$$

Table 8.2   ANOVA table.

Source of variation	Sum of squares	Degrees of freedom	Mean square	F ratio	p value
Regressor x	27.68	1	27.68	27.18	0.002
Residual	6.16	6	1.02		
Total	33.82	7			

For the sake of simplicity, we will not make any distinction between control and noise in this example. The variables x_1 and x_2 will be considered just as variables in general. Since we now are dealing with a continuous function, we need to exchange the sums in traditional ANOVA with integrals,

$$SS_{TOT} = \int_0^1 \int_0^1 \left(g(x_1, x_2) - \bar{g}\right)^2 dx_1 dx_2$$

where the 'average' value of g is given by

$$\bar{g} = \frac{\int_0^1 \int_0^1 g(x_1, x_2) dx_1 dx_2}{\text{Area}}.$$

In this example, the area is 1. This simplicity is actually the reason for choosing to work with $0 \leq x_i \leq 1$.

Define

$$g_0 = \int_0^1 \int_0^1 g(x_1, x_2) dx_1 dx_2 = \int_0^1 \int_0^1 \left(2x_1^2 + x_1 x_2\right) dx_1 dx_2 = \frac{11}{12}.$$

Since the 'area' is 1 in this example,

$$g_0 = \bar{g}.$$

We obtain

$$SS_{TOT} = \int_0^1 \int_0^1 \left(g(x_1, x_2) - g_0\right)^2 dx_1 dx_2$$

$$= \int_0^1 \int_0^1 \left(g(x_1, x_2)\right)^2 dx_1 dx_2 - g_0^2.$$

The marginal functions are obtained by integrating over all 'other' variables. We obtain

$$g_0 + g_1(x_1) = \int_0^1 g(x_1, x_2) dx_2 = \int_0^1 \left(2x_1^2 + x_1 x_2\right) dx_2 = 2x_1^2 + \frac{x_1}{2}$$

$$g_0 + g_2(x_2) = \int_0^1 g(x_1, x_2) dx_1 = \int_0^1 \left(2x_1^2 + x_1 x_2\right) dx_1 = \frac{2}{3} + \frac{x_2}{2}.$$

A third marginal function is $g_{12}(x_1, x_2)$. It represents the interaction between x_1 and x_2 and is implicitly given by

$$g_0 + g_1(x_1) + g_2(x_2) + g_{12}(x_1, x_2) = g(x_1, x_2) = 2x_1^2 + x_1 x_2.$$

In general for two variables,

$$g_0 + g_i(x_i) + g_j(x_j) + g_{ij}(x_i, x_j)$$
$$= \int g(x_1, \ldots, x_n) dx_i \cdots dx_{i-1} dx_{i+1} \cdots dx_{j-1} dx_{j+1} \cdots dx_n \qquad (8.9)$$

but since there are only two variables in total in our example, there is nothing left to integrate over in the last row of Equation (8.9). Thus we have

$$g_1(x_1) = 2x_1^2 + \frac{1}{2}x_1 - g_0 = 2x_1^2 + \frac{1}{2}x_1 - \frac{11}{12}$$
$$g_2(x_2) = \frac{2}{3} + \frac{1}{2}x_2 - g_0 = \frac{1}{2}x_2 - \frac{1}{4}$$

and

$$g_{12}(x_1, x_2) = g\left(x_1, x_2\right) - g_1 - g_2 - g_0 = x_1 x_2 - \frac{1}{2}x_1 - \frac{1}{2}x_2 + \frac{1}{4}.$$

These functions are illustrated graphically in Figures 8.9 and 8.10.

We are now prepared to divide SS_{TOT} into its components,

$$SS_{TOT} = \int_0^1 \int_0^1 g_1^2(x_1) dx_1 dx_2 + \int_0^1 \int_0^1 g_2^2(x_2) dx_1 dx_2 + \int_0^1 \int_0^1 g_{12}^2(x_1, x_2) dx_1 dx_2$$
$$= SS_1 + SS_2 + SS_{12},$$

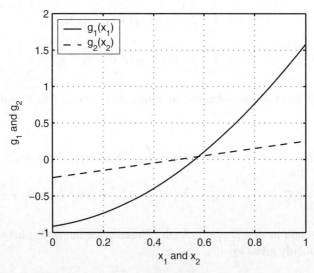

Figure 8.9   The marginal functions g_1 and g_2.

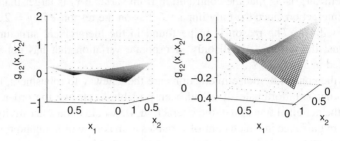

Figure 8.10   The marginal function g_{12} in two different scales, one that is the same as in Figure 8.9 plus one more adapted to this very function.

where SS_{12} is the interaction term. Numerically, we obtain

$$SS_{TOT} = 0.5708$$

$$SS_1 = \int_0^1 \int_0^1 g_1^2(x_1) dx_1 dx_2$$

$$= \int_0^1 \left(2x_1^2 + \frac{1}{2}x_1 - \frac{13}{11}\right)^2 dx_1 dx_2 = 0.5431$$

$$SS_2 = \int_0^1 \int_0^1 g_2^2(x_2) dx_1 dx_2 = 0.0208$$

and

$$SS_{12} = \int_0^1 g_{12}^2(x_1, x_2) dx_1 dx_2 = 0.0069.$$

8.3.3   Sensitivity indices

Besides the raw values from the sum of squares, there are other ways to express the impact of different variables. For example, the ratio

$$S_1 = \frac{SS_1}{SS_{TOT}} = \frac{0.5431}{0.5708} = 95\%,$$

which is called the global sensitivity index, or the Sobol index, for variable x_1. It can be interpreted as the percentage contribution to the total variation from this variable.

In the same manner, the global sensitivity index for x_2 and for the interaction are obtained from

$$S_2 = \frac{0.0208}{0.5708} = 3.6\%$$

$$S_{12} = \frac{0.0069}{0.5708} = 1.2\%.$$

It may seem surprising that the contribution from variable x_2 is larger than the one from the interaction since the only impact x_2 has on the response in $g = 2x_1^2 + x_1x_2$ is jointly with x_1. The reason is to be found in the hierarchical structure of the calculations; the term SS_{12} will only incorporate variation that has not yet been encountered in SS_1 and SS_2.

Since the control-by-noise interactions are the gateway to robustness, this hierarchical order is unfortunate. If it had been possible to change the order and first calculate the marginal function of the interaction, we would obtain sensitivity indices giving more guidance in which control variables to make use of to obtain robustness.

Besides the Sobol index S_i, there is another one that is just as useful. It is called the total sensitivity index (Homma and Saltelli, 1996). Since a variable like x_1 does not only contribute to the total variation on its own but also in interaction with other variables, there is sometimes a need to measure this total contribution. The total sensitivity index is

$$ST_i = 1 - S_{-i}$$

where S_{-i} is the sum of all Sobol indices not including the ith term. In our example, we have

$$ST_1 = 1 - S_2 = 1 - \frac{0.0208}{0.5708} = 96.4\%$$

and

$$ST_2 = 1 - S_1 = 1 - \frac{0.5431}{0.5708} = 5\%.$$

Example 8.8   Consider the function

$$g\left(x_1, x_2, z\right) = 5 - 7z - 6.3x_1z + 12x_2.$$

The marginal functions of the main variables are given by

$$g_0 = \int_0^1 \int_0^1 \int_0^1 g\,dx_1\,dx_2\,dz = 5.925$$
$$g_0 + g_{x_1} = 7.5 - 3.15x_1$$
$$g_0 + g_{x_2} = 12x_2 - 0.075$$
$$g_0 + g_z = 11 - 10.15z.$$

The marginal functions for the main terms are

$$g_{x_1}(x_1) = 1.575 - 3.15x_1$$
$$g_{x_2}(x_2) = 12x_2 - 6$$
$$g_z(z) = 5.075 - 10.15z.$$

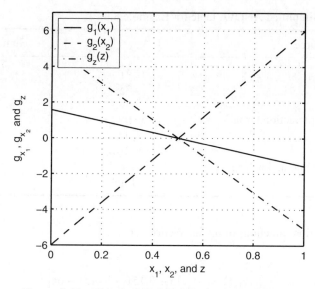

Figure 8.11   The marginal functions of Example 8.8.

These marginal functions are illustrated in Figure 8.11. It is worth noticing that x_2 is the biggest contributor to the variation, which is good news since it indicates that we have a control factor that at least can be used to tune to the target. However, the variation stemming from z is almost as large, so taking actions to reduce the effect by using robust design seems to be a useful approach.

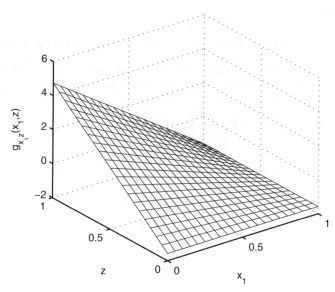

Figure 8.12   The marginal function of the x_1 by z interaction of Example 8.8.

Table 8.3   Simplified ANOVA table for Example 8.8.

Variation source		Sum of squares
Main terms	x_1	0.67
	x_2	12.00
	z	8.59
Two-factor interaction terms	x_1 by x_2	0.00
	x_1 by z	8.99
	x_2 by z	0.00
Three-factor interaction terms	x_1 by x_2 by z	0.00
	Total	30.25

The marginal functions of the interactions are

$$g_{x_1 x_2}(x_1, x_2) = 0$$
$$g_{x_1 z}(x_1, z) = -1.575 + 3.15x_1 + 3.15z - 6x_1 z$$
$$g_{x_2 z}(x_2, z) = 0$$
$$g_{x_1 x_2 z}(x_1, x_2, z) = 0.$$

This is sketched in Figure 8.12, where it is obvious that $x_1 = 1$ should be selected to obtain robustness against the noise z. The sums of squares are given in the ANOVA Table 8.3.

Exercises

8.1   Consider Example 8.2. Find the value x that gives type III optimal robustness.

8.2   Consider Example 8.8. Suppose that the target value is $g = 10$. What values should the control variables x_1 and x_2 have in order to both be robust against the noise z and aim at the target?

8.3   (Example 8.5). Find the Pareto optimal set of

$$y_1 = 10 + 1.7z + 1.7x_1 z - 0.3x_2 z, \qquad -1 \le x_1, x_2 \le 1$$
$$y_2 = 3 + 0.8z - 1.3x_1 z + 0.9x_2 z, \qquad -1 \le x_1, x_2 \le 1$$

with respect to type IV optimal robustness, subject to

$$u_1^2 \le 5$$
$$u_2^2 \le 1$$
$$E[y_1] \ge 10.5$$

under the assumption that $E[z] = 0$.

8.4 Use Taylor expansions of $g(x, z)$ around $z = E[Z]$ to discuss Equation (8.7). Apply this to Equation (8.2) and comment upon the conclusions.

References

Homma T and Saltelli A (1996) Importance measures in global sensitivity analysis of nonlinear models. *Reliability Engineering and System Safety*, **52**, 1–17.

Myers RH, Khuri AI, and Vining G (1992) Response surface alternatives to the Taguchi robust parameter design approach. *The American Statistician*, **46**, 131–139.

Wu Y and Wu A (2000) *Taguchi methods for robust design*. ASME Press, Fairfield, New Jersey.

9

Design and analysis of computer experiments

To design an experiment that does not take place in the physical world but in a virtual or simulated environment created with mathematical models offers both difficulties and opportunities of a specific kind and nature. There are several issues that must be addressed when an experiment in the physical world is planned that are irrelevant if the experiment takes place inside a computer. Take replicates, for example. It would just be a waste of time to have several replicates since it is known beforehand that identical results will be achieved for each one. On the other hand, new difficulties are faced with computer experiments. One of them is the quality of the mathematical model and how it should be explored, expressed, and judged in statistical terms. The fact that random variation is not present in an obvious way makes the concept of statistical significance somewhat vague and the statistical interpretation really challenging.

Design and analysis of computer experiments has been on the agenda at least since 1989 with the publication of some disseminating articles by Sacks *et al.* (1989a, 1989b). The methodology has found its way into the software packages for finite element analysis and has therefore quickly reached a fairly high level of usage in industrial applications. The books by Fang *et al.* (2006) and Santner *et al.* (2003) both give a more solid treatment of the area than could possibly be captured in this chapter.

There are many types of computer experiments – far too many to cover in one single chapter, or even in a single book. We need to focus on something, and the type of experiments in mind while writing this chapter has been finite element analysis and related topics. Nevertheless, substantial parts of the chapter are applicable for a much broader range.

Statistical Robust Design: An Industrial Perspective, First Edition. Magnus Arnér.
© 2014 John Wiley & Sons, Ltd. Published 2014 by John Wiley & Sons, Ltd.
Companion website: www.wiley.com/go/robust

Robust design based on virtual experimentation has one great advantage. It can be performed at an early stage, a stage when the engineer still has a substantial amount of freedom.

Even if the topic of this book is robust design, it does not make sense to discuss simulation experiments for robust design without discussing the specific problems and opportunities of design of computer experiments in general. Thus, this chapter differs from the rest of the book in the sense that it is not primarily about robust design. Robust design will just be a special case in a bigger picture.

9.1 Overview of computer experiments

A first course in DOE usually has an emphasis on factorial designs, primarily two-level designs. Concepts such as randomization, replicates, and blocking are central elements. The aim of such a course is to bring the student the knowledge that is important for designing experiments in the physical world. For virtual experiments, on the other hand, these concepts are not necessarily of much interest. Assume, for example, that a finite element code for some mechanical problem is written and that an experiment is designed in order to learn how some factors, say some material or geometry parameters, affect the result. To have several replicates would then be pointless. If one calculation is repeated, the same result will simply be obtained once again. For the same reason, randomization is uninteresting. In addition, it turns out that two-level factorial designs are relatively inefficient. It is better to have some kind of 'space-filling design' in order to explore the entire design space. Design of virtual experiments will thus differ in many ways from physical experiments, but will not be of a completely different nature.

The problems that are addressed with virtual experiments are of a relatively wide range. For example, the purpose might be

- to understand which factors have the biggest influence on the response;
- to build a mathematical model;
- robust design;
- optimization, including multi target and multi disciplinary optimization;
- validation of the computer model ('math model validation');
- tuning, or calibration, of the computer model.

The critical reader is probably surprised by the two first items on this list. Have these questions not been answered already? If there is a virtual model, for example a finite element model, then there is already a mathematical model, otherwise it would not have been possible to write the code. Thus, it is already known which factors have the biggest effect and thus designing such experiments is pointless.

These objections are certainly correct but yet overlook something, namely the model complexity. If the engineers who have written the code are asked which factors have the biggest impact, the response is usually that they have no idea. If a question about robust design or optimization is asked, the answer is often that the calculations are so time consuming that they cannot perform them even if they wanted to. They are simply in need of a simpler model, a meta model, or a surrogate model, as it is also called. Such a model is what is meant in the list above.

The steps taken in computer experiments are not exactly the same for every problem listed above. For example, the steps taken to validate a mathematical model are slightly different from the steps of robust design. However, they have all so much in common that an example of one of them can provide a general understanding. To illustrate this, let us take one example, robust design.

9.1.1   Robust design

A possible set of steps in robust design for computer experiments can be

   (i) Build the original mathematical model.

  (ii) Set up the robust design problem (response, signal, noise, and control factors).

 (iii) Identify the design space (for the control factors) and variation space (for the noise factors).

  (iv) Set up a so-called space filling experimental design to explore the space identified in step 3.

   (v) Run the calculations of the experimental design.

  (vi) Build a response surface.

 (vii) Use the response surface and optimize the objective.

(viii) Confirmation run.

Build the math model

The first one of these steps, to build the mathematical model, is not specific to DOE. The intended model is a finite element model. In principle, the way to build the model is the same as always used in finite element analysis. However, one major consideration must be taken to make it possible to use the model for factorial experiments. It must be possible to change the values (levels) of factors easily, and this will have some consequences for the model building. Things that can both be changed by the engineer and modelled mathematically include material parameters like the elasticity modulus and material density, and geometry parameters are typical factors in the experiment. Material parameters are mostly easy to change in the model. The problem is the geometry parameters. A change of value of some geometry parameters should not require remeshing. A parameterized geometry is useful for such a purpose, and the commercial software packages usually have good opportunities for parameterizing the geometry.

Set up the robust design problem

In order to optimize something or to build a response surface, the response must be numerical. If the response is not just numerical but in addition takes a single value it is much easier to use than if it is a function. This is sometimes a problem. A response function has to be summarized in a few scalars to be able to proceed.

Only noise factors that can be modelled mathematically can be taken into consideration. This typically includes

- variation around the nominal of material parameters;
- variation around the nominal of geometry parameters;
- loads and boundary conditions.

For material parameters, random variation between units (e.g. package to package or car to car) or within a unit (a disturbance from node to node within one and the same unit) may be considered. Within-unit variation can typically be sheet thickness variation or elasticity modulus variation over a carton board or a metal sheet. Modelling variation of this kind require usage of more advanced statistical techniques, such as random fields.

Noise factors are often variation around the nominal. Thus, several factors are both control and noise. The nominal value is a control factor and the random variation around the nominal is a noise factor.

In order to select the levels of the factors, both control and noise, it must first be known what space of parameter values to explore. For the control factors, it will be the space within which the engineer has the freedom to select the value. This is the 'design space'. For the noise factors it is the space within which the noise is supposed to take its values. If a distinction between the design space and this second one is needed, the latter will be called the 'variation space', but mostly both of them are referred to as the 'design space'. It is certainly difficult to know what the design and variation space will be. Especially the variation space has a tendency to be more or less unknown. However, there are some sources of knowledge, at least if a similar type of product has been produced in the past. Sources of information about the variation space might include data from manufacturing and data from suppliers about 'quality classes'. Even tolerance limits and capability requirements might give valuable information, but only occasionally since they are usually not available at this stage in the product development – or at least they should not be since tolerancing should come after rather than before robust design activities.

The experimental array

In this stage of the experimental design no difference is made between noise and control factors. They are just considered as factors in general. It is not until later that this distinction is made. This double nature of the noise factor is one of the largest conceptual difficulties to overcome for the practitioner.

The experiment can be divided into two steps, one screening design followed by a so-called space filling design. The type of screening design used in computer

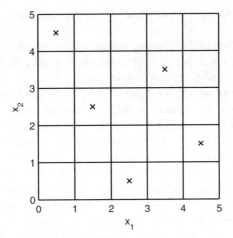

Figure 9.1   A Latin square for the space [0   5] × [0   5]. *Each row and each column is represented once.*

experiments is often the same as the ones used in physical testing. Two-level designs that are strongly reduced or even saturated, such as 2^{7-4} designs, as well as Plackett–Burman designs, are typical examples, but the main experiment is usually different from those of physical experimentation. Various types of 'space filling designs' are useful. The most common type of such an array is the Latin hypercube, which is described as follows.

Suppose there are two factors, x_1 and x_2. The rectangular space in Figure 9.1 will be investigated. In a Latin square, the space is divided into rows and columns, and the runs in the experiment are spread out in such a way that each row and each column is represented once.

There are other space filling designs apart from the Latin hypercube. Some of them may very well have better space filling properties than a Latin hypercube.

Response surfaces

The simulation results of the runs from the space filling design can be used to estimate a response surface. The primary reason for a response surface is that the time it takes to run the original finite element model is long, so that a response surface is created and used as a meta model or surrogate model, i.e. a model of the model. Once this meta model is in place, it is used as if it had been the true model.

The simplest type of a response surface is a linear regression model, but for the purpose of using it as a surrogate model, it would be insufficient. Something more elaborate is needed. Some common approaches are

- polynomial regression;
- local least squares;
- kriging.

In Section 9.3, these ways to estimate the response surface are treated in more detail.

During the creation of the response surface no distinction is made between noise and control factors, which in turn is the reason not to make such a distinction when setting up the space filling experimental design.

Once the response surface is estimated, it has to be validated. Is it really a good enough substitute for the original model? The typical way to check this is cross validation, first and foremost one-out-at-a-time cross validation (see Hjorth, 1994, for details or Fang et al., 2006 for its application to computer experiments). If the accuracy of the response surface is unsatisfactory, there are techniques that can be used to identify points in the design space from which to run additional finite element calculations. Jumping back and forth between the experimental array, the response surface and the optimization is also useful. The closer to the optimum, the more concerned we are about the quality of the response surface, so during the course of the optimization, fill-in points in the experimental design may very well be identified and the response surface improved.

Optimization for robustness

Once the response surface is in place, it is time to optimize for robustness. It is at this stage that the distinction between noise and control factors is made, the distinction that was left aside when the experimental array was set up and the response surface formed.

Questions that need to be addressed include what type of optimization algorithm to use. In commercial software packages available for the purpose, the list of such algorithms is long, not to say breathtaking. Another and even more delicate issue concerns what to optimize. Two options that sometimes can be useful is the loss function,

$$L(y) = K(y - m)^2$$

where m is the target value, and the sensitivity to noise,

$$u^2 = \left(\frac{\partial y}{\partial z}\right)^2. \tag{9.1}$$

For the criterion of Equation (9.1) there are a number of varieties, as discussed in Chapter 8. From the point of view of robust design, this second criterion is often the most natural one.

There are several other possibilities, as will be discussed in Section 9.4.

Confirmation run

Since the optimization is performed on a meta model, it needs to be confirmed that the original model, the finite element model, gives the same, or at least a similar, result in the optimal solution. This task might seem easier than it is. The average value can easily be checked, but there can be a problem with the sensitivity. It may require several runs, which could be costly from a time perspective.

9.2   Experimental arrays for computer experiments

Two very different kinds of experimental designs are used for virtual experiments. One is screening designs, whose purpose is to reduce the number of factors to something manageable. These designs are usually fractional factorial designs with the factors on two levels, that is the same kind of designs used for any kind of screening. The other type is space filling designs.

9.2.1   Screening designs

The number of potential factors in simulation based experiments (or any other kind of designed experiments) is sometimes so large that it is necessary to perform testing in two steps, first by screen out the most important ones and then by performing the main experiment. For such screening experiments, two-level factorial designs are commonly used. There are of course some risks attached to such an approach. Specifically, factors with large interaction effects are not always identified as important. Since interaction is the key to robust design, a screening experiment of this kind may jeopardize the chance to achieve robustness. Nevertheless, screening is often a necessity. The alternative is not to explore all parameters. For time cost limitations, it is rather not to run any experiment at all.

Fully saturated factorial designs on two levels are commonly used in screening (a design is saturated if there are k factors and $k + 1$ runs). Consider the saturated design in Table 9.1. In this design, factor A is confounded with several second order interactions,

$$A = -BD = -CE = -FG.$$

Table 9.1   A saturated factorial design.

Run	A	B	C	D	E	F	G
1	-1	-1	-1	-1	-1	-1	-1
2	1	-1	-1	1	1	-1	1
3	-1	1	-1	1	-1	1	1
4	1	1	-1	-1	1	1	-1
5	-1	-1	1	-1	1	1	1
6	1	-1	1	1	-1	1	-1
7	-1	1	1	1	1	-1	-1
8	1	1	1	-1	-1	-1	1

There are of course situations when a saturated design of this kind leads the experimentalist in the wrong direction. Assume, for example, that the true model is

$$y = 1.7 + 0.5x_B - 0.6x_D + 0.4x_{BD} + \varepsilon, \quad \varepsilon \text{ is } N(0, \sigma).$$

If the standard deviation σ is small enough, then the screening design of Table 9.1 has a tendency to point out factors A, B, and D as the vital ones to be included in the main experiment. Thus, the screening will erroneously identify factor A as important. A way to avoid this but still keep the experimental design as small as possible would be to use a Plackett–Burman design or a Taguchi type orthogonal array (see Chapter 11).

A screening experiment that points out the few vital factors as well as some additional factors may be acceptable. The consequence is that the main experiment will be larger than needed. It is a bigger issue if some important factors are dismissed as unimportant. This may be the fact if the true model is

$$y = 1.7 + 0.4x_A + 0.5x_B - 0.6x_D + 0.4x_{BD} + \varepsilon, \quad \varepsilon \text{ is } N(0, \sigma).$$

The effects of factor A and the interaction BD will cancel out each other, making us believe that there is no effect of factor A. The remedy would be to increase the size of the experimental design to a 2^{7-3} array. The resolution will then be IV and no main effect will be confounded with a second order interaction. Which one of these two should be chosen, a 2^{7-4} design (the saturated one) or a 2^{7-3} design, depends on a number of things. The risk that one interaction, or a combination of interactions, will cancel out the effect of a main factor in this manner is mostly small, which speaks in favour of the smaller array. Speaking in favour of the larger design, apart from its resolution, is the fact that it contains more runs, decreasing the impact that the 'random' term ε has on the conclusions.

The last one of the arguments, pure size, can in physical experiments be approached in another way, namely to make use of a saturated design with replicates. However, for virtual experiments such an approach will be useless.

In some situations, a saturated design will fail completely if used as a screening design. One situation is when there is a second order effect but no first order effect, as in

$$y = 1.7 + 0.4x_A^2 + 0.5x_B - 0.6x_D + \varepsilon, \quad \varepsilon \text{ is } N(0, \sigma). \tag{9.2}$$

Another would be if there are factors that are important in interactions but not on their own, as factor B in

$$y = 1.7 - 0.6x_D + 0.4x_{BD} + \varepsilon, \quad \varepsilon \text{ is } N(0, \sigma). \tag{9.3}$$

For the second order model (Equation 9.2), an experimental design on two levels will always fail as more levels are needed. For screening, a Taguchi type orthogonal array such as L_{18} may be appropriate (see Chapter 11). For the model of Equation (9.3), the interaction will require a higher resolution of the experiment. From a robust design point of view, the issue of the model in Equation (9.3) is of great importance since the very idea of robust design is to exploit interaction effects. This speaks in favour of a higher resolution of screening designs in robust design than ordinary DOE.

The principle of hierarchy

Several software packages and textbooks on DOE emphasize 'the principle of hierarchy'. The meaning of this principle is that unless a factor is important of its own, it will not be important in an interaction. In the software Minitab, for example, it is not even possible to include the interaction BD in the model unless both factors B and D are included. There is of course some rationale behind this principle; higher order interactions are usually smaller in size than main effects and experience has shown that factors included in the vital interactions are mostly of substantial importance on their own. Having this in mind, none of the models in Equations (9.2) and, especially, (9.3) is a major problem since it is unlikely that they exist. However, the principle of hierarchy may be overestimated. For example, it assumes that factors have been selected in an intelligent way. The conclusion is therefore that, despite the principle of hierarchy, the situations pointed out in the models of Equations (9.2) and (9.3) are really problems in screening. This problem is bigger in robust design where the main interest is interactions rather than main effects.

9.2.2   Space filling designs

Since replicates are of no value in computer experiments the factor levels can be changed more freely than in physical testing. Suppose, for example, that we can afford to have 16 runs. Rather than having three factors on two levels each and two replicates, something else can be done. One alternative is to have four factors on two levels each, a 2^4 design. Another is to have two factors on four levels each, a 4^2 design. None of these is really appealing. To increase the number of factors, as in the first alternative, will not allow us to include polynomial terms and the outcome of the experiment might be too poor to build a useful response surface. The other alternative, to increase the number of levels, will just give information about two factors, not three, as in the original problem. Thus, two-level factorial designs are not the primary choice in virtual experiments (except for screening).

The goal is to explore the behaviour within the design space. A space filling design is well fitted for this purpose. We have already looked into one example of such designs, Latin hypercubes.

Properties of a space filling design

For traditional experimental designs, there is a well developed optimality theory. There is a wide range of alphabetical optimality criteria, among which A optimality and D optimality are the ones most widely used. An underlying assumption for all these criteria is that it is known in advance which parameters to estimate, and the optimality criteria deal with the confidence regions of the parameter estimates. For example, assume that a two-level factorial design is used to estimate the parameters in the model

$$y = \beta_0 + \beta_A x_A + \beta_B x_B + \beta_C x_C + \beta_D x_D + \cdots + \beta_{ABCD} x_{ABCD} + \varepsilon.$$

A full factorial design with two replicates would mean an experiment with $2 \times 2^4 = 32$ runs. However, if the experimenter only can afford 20 runs, which ones among the 32 should be run if there is a specific interest in the parameters β_A, β_B, β_C, and β_{AC}? A design is a D optimal design if the simultaneous confidence region of these four estimates is minimized, that is a four-dimensional hyper-volume. Thus, in order to get a D optimal design, we need to know in advance what we should be looking for (in this case first and foremost the four coefficients).

In contrast to these traditional optimality criteria where the focus is on the way in which the experimental results will be used, the focus in computer experiments is on the design matrix X. In a space filling design X, the experimental points should be well spread out over the design space, so this space is well represented. There are primarily two ways to express this, namely the distance between experimental points and the entropy:

- Maximum distance: the experimental design that maximizes the minimum distance between two neighbouring experimental points.

- Maximum entropy: a mathematical concept to measure how 'well spread out' in the space that the points are.

The details of these two concepts will not be discussed in this book. They are well treated in the books by Santner *et al.* (2003) and Fang *et al.* (2006).

There is a difference in optimality concepts since in computer experiments of this kind the goal is to create a surrogate model that the engineer can make use of for simulation and to find answers to a lot of questions, some of which may not even have been formulated on the occasion when the experimental design was defined. That is at least how it can be motivated, and usage of these criteria has become more or less standard in software packages for computer experiments.

In practice, these optimality criteria can be used as a guide to finding a good experimental design, but it makes little or no sense to actually find the optimal experimental design. When the experiment is run, some runs will fail. Some will fail because there are convergence problems, others due to technical or numerical problems. Some can be re-run to obtain a result, some cannot. To have an experimental design that truly is optimal when it cannot be run in full does not make sense. It is more important that an experimental design for computer experiments has reasonably good space filling properties and is robust against lost runs.

Various space filling designs

There are many different families of space filling designs that are implemented in commercial software packages. Among them we find

- Latin hypercubes;

- Hammersley designs;

- stratified random sampling.

In this book we will just consider the first one of these. Readers interested in the others are referred to the manuals from software providers.

9.2.3   Latin hypercubes

Even if the value of finding the optimal experimental design in terms of the maximum distance is limited, it is worth spending some effort finding an experimental design that has fairly good distance properties. We will see how this can be accomplished for a Latin hypercube.

Consider the Latin square

$$X = \begin{pmatrix} 1 & 1 \\ 2 & 2 \\ 3 & 3 \\ 4 & 4 \\ 5 & 5 \end{pmatrix}$$

(Figure 9.2). It is certainly a Latin square since each row and each column is represented once. However, it is not space filling. The algorithm to improve the space filling property in most software packages is very simple and based on random changes.

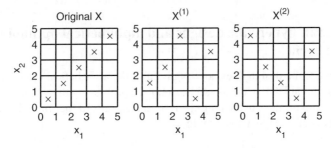

Figure 9.2   Three Latin squares with different distance properties.

One such algorithm to maximize the minimum distance in a space filling design is to write all columns but the first in a random order, for example

$$X^{(1)} = \begin{pmatrix} 1 & 2 \\ 2 & 3 \\ 3 & 5 \\ 4 & 1 \\ 5 & 4 \end{pmatrix}.$$

The size of the smallest distance d and the number J of pairs of this distance in the new matrix is calculated. In the case of $X^{(1)}$, we have $d = \sqrt{2}$ and $J = 1$. Then find the pair of points with the smallest distance. Randomly pick one of the rows in the pair and one row that is not in the pair and swap the two rows. A new candidate design $X^{(2)}$ is obtained. Once again, calculate d and J for the new candidate. If d is increased or if J is reduced with an equal distance d, then update the design matrix X to the new design. Otherwise, stay at the old design matrix and take another pair to be swapped. For $X^{(1)}$, rows 1 and 2 constitute the pair with the smallest distance. For one of the rows in this pair, say row 1, and a randomly selected row, say row 3, the values are swapped. The new candidate design is

$$X^{(2)} = \begin{pmatrix} 1 & 5 \\ 2 & 3 \\ 3 & 2 \\ 4 & 1 \\ 5 & 4 \end{pmatrix}.$$

The smallest distance $d = \sqrt{2}$ is unchanged but there are two pairs of points with this distance, $J = 2$. Since this is larger than in $X^{(1)}$, the new candidate is discarded and another two rows of $X^{(1)}$ are picked in the same manner. The row swapping continues either until some criteria is reached or until a prespecified number of row swaps are performed.

For a Latin square with five rows and five columns of this kind, placing the points like the knight's move on a chessboard (Figure 9.1) is actually is the maximum distance design.

9.2.4   Latin hypercube designs and alphabetical optimality criteria

A Latin hypercube design does not usually fulfil the alphabetical optimality criteria used in traditional DOE. Consider, for example, a problem with two factors, A and B, that can both take values in the range $[-1, 1]$ and can afford four runs. The D optimal design, that is the design X that maximizes

$$\det(X'X),$$

is

$$X = \begin{pmatrix} -1 & -1 \\ 1 & -1 \\ -1 & 1 \\ 1 & 1 \end{pmatrix} \tag{9.4}$$

giving $\det(X'X) = 16$. On the other hand, the maximum distance Latin square for the same factor space is

$$V = \begin{pmatrix} -1 & 1/3 \\ -1/3 & -1 \\ 1/3 & 1 \\ 1 & -1/3 \end{pmatrix}$$

for which $\det(V'V) \approx 4.9$, which is smaller than 16.

The practical meaning of D optimality is that the simultaneous confidence region for the parameter estimates is minimized. Thus, a traditional design, which is D optimal, will give more information about the parameters of the model. However, it requires something, namely that the model structure is known in advance. The design matrix X of Equation (9.4) will, for example, minimize the simultaneous confidence region of the parameters μ, β_A, β_B, and β_{AB} in the model

$$y = \mu + \beta_A x_A + \beta_B x_B + \beta_{AB} x_A x_B + \varepsilon,$$

given the design space $-1 \leq x_i \leq 1$. One characteristic in DOE for computer experiments is that the structural model form is unknown. We want to create information that can give answers to questions not yet formulated. For this purpose, Latin hypercubes have properties that D optimality and similar criteria do not posses.

9.3   Response surfaces

The type of response surface most widely used in industrial practice is probably kriging. The reason could be that kriging can fit a surface that actually goes through each calculated point. This contrasts with every type of regression, where the best fit line, from a least squares point of view, does not pass through all points. In computer experiments this is appealing. The reason is the lack of true randomness. In the calculated points, something substantial is known. It is only between the points that there is an uncertainty, and the further away from the calculated points, the larger the uncertainty. Kriging makes it possible to describe this uncertainty in statistical terms, making use of confidence bands.

Even though it is likely that kriging is the most widely used technique for response surfaces in computer experiments, it will more or less be left aside in this discussion. Polynomial regression is often a good choice, but it is well treated in most basic textbooks on regression and will not be described here. We will confine our treatment of response surfaces to local least squares.

9.3.1   Local least squares

In ordinary least squares in regression, the coefficients β are estimated by

$$\hat{\beta} = \left(X'X\right)^{-1}\left(X'Y\right).$$

Thus, every observation y has equal weight. In weighted least squares, some observations have more weight than others,

$$\hat{\beta} = \left(X'WX\right)^{-1}\left(X'WY\right),$$

where W is a diagonal matrix of weights. Weighted least squares is a useful method, for example, when the uncertainty or variation that comes with the observation differs from observation to observation. In physical experiments this situation may be present if the measurement equipment that has been used is not the same for all observations. In virtual experiments, the observations may have been collected using two different models, one with a finer and one with a coarser mesh grid. Weighting is a way to pay more attention to the high quality data.

Since all observations have equal weight in ordinary regression, the value of the regression line in a point x_0 is affected by remote points as well as close neighbours, all of them with equal weight. This may be a problem. Locally weighted least squares is a way to address the problem. It is a special case of weighted least squares where nearby points have a larger influence than remote points. The estimated coefficients

$$\hat{\beta}(x) = \left(X'W(x)X\right)^{-1}\left(X'W(x)Y\right)$$

become functions of the value of x (which is a $p + 1$ dimensional vector if there are p regressors).

The weighting can be performed in many different ways. One way is to have a rectangular window, so that all observations falling within the rectangle are included with its full weight and everything falling outside of the rectangle is removed,

$$w_{ii}(x) = \begin{cases} 1, & \text{if } -b < x - x_i < b \\ 0, & \text{otherwise} \end{cases}$$

where w_{ii} is the ith diagonal element of W. For various reasons, such a window is not very efficient. Another opportunity is a Gaussian window,

$$w_{ii}(x) = \exp\left(-\frac{(x - x_i)^2}{2\theta^2}\right)$$

where θ can be interpreted as a kind of 'window width'. With these weights, all observations are taken into account but nearby points have a higher weight.

The problem with local least squares, no matter whether the weighting window is Gaussian or not, is how the parameter θ in the weights should be set. It is not just to estimate it with least squares in a straightforward fashion since with a small enough value of θ the regression curve (or surface) will pass through all observations and give a model that perfectly fits the available data but is useless for the prediction of new observations. It will simply be linear interpolation between the closest observations. One way to approach the estimation problem is cross validation, excluding one observation at a time in order to find the value of θ that minimizes the prediction error sum of squares (PRESS) value. Another opportunity is to use the covariance function of the data to obtain the weights (this is the approach used in kriging).

Example 9.1 Suppose that we have the function

$$y = 2x^2 \sin(x). \tag{9.5}$$

We will fit a meta model (response surface) to this function by using local least squares with Gaussian weights and window width $\theta = 1$,

$$w_{ii} = \exp\left(-\frac{(x - x_i)^2}{2}\right).$$

Thus, the window width is assigned to a particular value rather than being estimated (this is explored further in Exercise 9.2). Suppose that the function of Equation (9.5) has been evaluated for 10 values of x, namely $X = 0, \ldots, 10$. An ordinary linear regression model will not fit the data very well (Figure 9.3). Polynomial regression is a better choice, but the order of the polynomial needs to be fairly high to give a good fit. In Figure 9.3, a fourth order polynomial is fitted to the data. Comparing that to the local least squares in the same figure, even a visual examination points out the local least squares as being better. It should be remembered that the parameter estimates obtained in the polynomial regression are optimal in a least squares sense. The parameter θ in the local least squares model was picked somewhat arbitrarily, so for the function of Equation (9.5) the local least squares method is likely to be better.

9.3.2   Kriging

The simplest kind of kriging model is

$$y(x) = \mu + v(x)$$

where μ is the overall mean of $y(x)$ and $v(x)$ is a process with mean $E[v(x)] = 0$ and covariance function

$$Cov\left(v(x_i), v(x_j)\right) = \sigma^2 R(x_i, x_j).$$

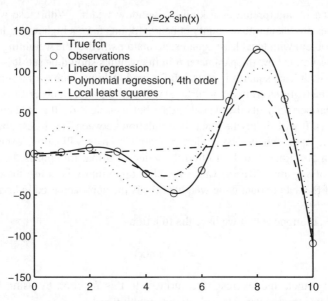

Figure 9.3   The function $y = 2x^2 \sin(x)$ *evaluated at 11 points and modelled using different types of response surfaces.*

This covariance function is assumed to have a functional form of some kind. A common form is the Gaussian covariance function,

$$R(x_i, x_j) = \sigma^2 \exp(-\theta(x_i - x_j)^2) = \sigma^2 r(x_i, x_j). \tag{9.6}$$

This reflects the basic idea of kriging, namely to have a functional form of the covariance function rather than, say, a prediction equation as in regression analysis. The problem is reduced to an estimation of unknown parameters in the covariance function (σ and θ in Equation 9.6). The covariance in Equation (9.6) depends only on the distance between the points x_i and x_j. In a slightly more complicated case, the correlation may depend on the direction,

$$r(x_i, x_j) = \exp(-\theta_k(x_{ik} - x_{jk})^2).$$

The resulting response surface can be written as

$$g(x) = \sum_{i=1}^{n} y_i r(x, x_i)$$

where n is the number of runs in the experimental design (number of available data points). Thus, the value of the response surface in the point x is a linear combination of all available data. Remote points will have less weight, close points a higher weight.

9.4   Optimization

Once the response surface is in place, it is time to pick up the distinction between noise and control parameters in order to find a robust solution. This solution can be found in many different ways. One is just to look at the graphical representation of the functional ANOVA. However, even if it is good that such a visual examination is a part of the analysis, it is mostly insufficient on its own. Optimization methods offer a more solid way to proceed. Such an approach is easily available using commercial software.

9.4.1   The objective function

The number of optimization algorithms available in commercial software for computer experiments is astounding. In the manuals to this software, however, there is little or nothing written on what to optimize. The problem is not that nothing is written about which forces, tensions, or temperatures to optimize. The missing part is how the problem should be formulated in order to optimize for robustness. There are several alternatives.

The various types of optimal robustness introduced in Chapter 8 are good choices for objective functions, or at least two of them, namely type IV and type V,

$$\min_x E\left[(u(x, Z))^2\right] = E\left[\left(\frac{\partial g(x, Z)}{\partial Z}\right)^2\right] \quad \text{(type IV robustness)}$$

or

$$\min_x Var\left[g(x, Z)\right] = \int (g(x, z) - E\left[g(x, z)\right])^2 f_Z(z)dz$$

$$\text{(type V robustness)}.$$

Both of these criteria have drawbacks. First, both involve integration. If the noise Z is multidimensional, this may be cumbersome and time consuming, even on the response surface. However, there is another drawback that is even more important. They both require the probability distribution $f_Z(z)$ of the noise Z to be known. Such knowledge is uncommon. There is seldom more than a rough idea of the mean value of Z that is available. Thus, a simplified version of type IV robustness is often more practical, one where the function $g(x, z)$ only is differentiated at the mean of Z. Thus, the objective function becomes

$$\min_x \left(u(x, \mu_z)\right)^2 = \left(\frac{\partial g(x, z)}{\partial z}\bigg|_{z=\mu_z}\right)^2. \tag{9.7}$$

The two-step approach of robust design, where the function is made as robust as possible in the first step and then the remaining control factors are used to tune to target, can also be used for computer experiments. It is often a good way to approach the problem, but there are alternatives. Minimizing the loss function is one, but not necessarily the best, alternative. Making use of subjective functions may be more appropriate. The problem can thus be formulated as

$$\min_x \ \left(u(x, \mu_z)\right)^2$$
$$\text{subject to } -a < g(x, \mu_z) - m < a$$

where m is the target value. Alternatively, it could be

$$\min \ \left(g(x, \mu_z) - m\right)^2$$
$$\text{subject to } \left(u(x, \mu_z)\right)^2 < b.$$

Another possibility is to make use of Pareto optimality in a somewhat unusual way, namely to treat the differentiated function u and the main function g separately and search the Pareto optimal solution for

$$\min_x \ \left(u(x, \mu_z)\right)^2$$
$$\min_x \ \left(g(x, \mu_z) - m\right)^2.$$

The problem with this approach is that there is usually not just one function g but several, and the number of objective functions in the multiobjective problem will increase by a factor of two in this way.

Many requirements are expressed in terms of probabilities. A lifetime requirement is one example. Typically, such a requirement may be of the type

$$P(T > 1000 \text{ h}) > 0.99$$

for a lifetime T or

$$P(1000 \text{ ml} < V < (1000 + a) \text{ ml}) > 0.99$$

for a filling volume V of a bottle. This can actually be used as an optimality criterion as well. The optimization problem for the lifetime may be stated as

$$\max_x \ \int_{1000}^{\infty} T(x, Z) f_Z(z) dz$$

and for the filling volume as

$$\max_x \ \int_{1000}^{1000+a} V(x, Z) f_Z(z) dz.$$

Thus, we have seen a number of possible objective functions for robust optimization on a response surface. My personal opinion is that the baseline should be the function of Equation (9.7), possibly combined with some subjective functions.

9.4.2   Analytical techniques or Monte Carlo

Noise can be handled in different ways in the optimization. We have already seen one of them, namely to differentiate the response with respect to the noise and then to minimize. However, the impact the noise has on the response can also be handled in another way. Once the response surface is in place, only the control factor levels are set to fixed values; then noise is generated at random from the distribution $f_Z(z)$. In Monte Carlo simulation a group of several random numbers are picked, not just once but many times. We can call each such time a 'shot'. For each fixed set of control factor levels, there may be hundreds of shots of the noise factors. This is then used to estimate the variance. The design space is then searched to find the point that is most robust to noise. This search can be carried out in several different ways. One is to make use of designed experiments on the response surface and look for dispersion effects. Sometimes it is possible to use optimization techniques. However, we will leave this for the next chapter where Monte Carlo methods are studied in more detail.

If we compare the two methods, we realize that the one we have studied so far is more or less based on Taylor expansions, first and foremost Taylor expansions of the first order. Such an approach will not give the exact solution since the Taylor expansion is only an approximation (compare Equation 8.7). A Monte Carlo approach will, on the other hand, give an exact solution if the number of shots goes to infinity and the probability distribution of the noise is known. The drawback is usually that this distribution is unknown. In addition, using Monte Carlo simulation requires a response surface to be available or the original model to be fast in terms of calculation time (see Chapter 10). Running Monte Carlo simulation on the original finite element model is usually far too time consuming to be practically possible.

Example 9.2   Consider the function

$$g(x, z) = 2.3 + 4(x + z) - (x + z)^8, \ 0 < x < 1.25$$

(Figure 9.4) where the noise z is a random variation around the nominal. The noise is $N(0, \sigma)$ where $\sigma = 0.1$, and the function should be maximized.

Let $v = x + z$. The function is maximized in the point where

$$\frac{\mathrm{d}g}{\mathrm{d}v} = 4v - 8v^7 = 0,$$

that is for $v = 0.5^{1/7} \approx 0.95$. Since the noise is a small disturbance around the nominal, its derivative is zero in the same point. Thus, with this approach, $x \approx 0.95$ is the optimal choice both with respect to maximization of the mean and robustness to noise. However, it is clear from Figure 9.4 that this does not hold since

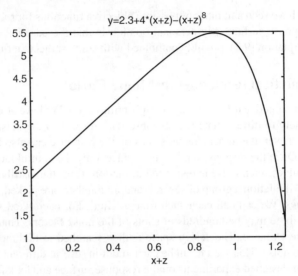

Figure 9.4   The function in Example 9.2.

the function is unsymmetrical. At this point the response is highly sensitive to the noise.

A Monte Carlo simulation of the same problem with 500 Monte Carlo shots for all values of x from zero to ten in steps of 0.012 units gives a different result (Figure 9.5). A few things are worth noticing. The empirical average reaches its maximum at $x = 0.87$, a smaller number than in the other approach, and the variance at $x = 0.85$.

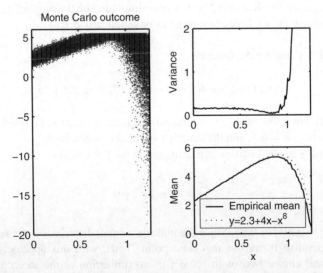

Figure 9.5   A Monte Carlo simulation of the function in Example 9.2.

The fact that the mean curve has shifted is interesting. This is a phenomenon that often is left unconsidered when the nominal value is set. Thus, in this example the Monte Carlo approach turned out to be better than the analytical one.

The same example can be approached using the matrix formulation of Equation (8.4). In order to minimize the transmitted variance

$$\left(\beta'_z + X_x B\right) \Sigma \left(\beta'_z + X_x B\right)'$$

we need the vectors and matrices β_z, X_x, B, and Σ. The three first are

$$\beta_z = (4, 0, 0, 0, 0, 0, 0, -1)'$$
$$X_x = \left(x, x^2, x^3, \ldots, x^8\right)$$

and

$$B = \begin{pmatrix} 0 & 0 & 0 & 0 & 0 & 0 & 8 & 0 \\ 0 & 0 & 0 & 0 & 0 & 28 & 0 & 0 \\ \vdots & \vdots & \vdots & \vdots & \vdots & \vdots & \vdots & \vdots \\ 8 & 0 & 0 & 0 & 0 & 0 & 0 & 0 \\ 0 & 0 & 0 & 0 & 0 & 0 & 0 & 0 \end{pmatrix}$$

where the coefficients in B are obtained through the evaluation of $(x + z)^8$. The matrix Σ is more of a problem since involves terms like $Cov(Z, Z^7)$ and $Var(Z^8)$, that is higher order moments of the normal distribution that are difficult to calculate.

Exercises

9.1 Assume that an experimentalist has a list of seven possible factors and is about to run a screening design to identify the most vital ones for further experimentation. Write a code in some software to generate data from the model

$$y = 1.7 + 0.5x_B - 0.6x_D + 0.4x_{BD} + \varepsilon, \quad \varepsilon \text{ is } N(0, \sigma)$$

where $\sigma = 0.1$. Make use of
(a) A two-level saturated experimental design, that is a 2^{7-4} design as in Table 9.1.
(b) A Plackett–Burman L_{12} design (Table 9.2) with four empty columns. Reflect on the differences in the conclusions.

9.2 Recall Example 9.1. Use cross validation by excluding one observation at a time from the data and estimate the optimal value of the window width θ by minimizing PRESS, the prediction error sum of squares.

Table 9.2   A Plackett–Burman design with 12 runs and four unused columns.

Run	A	B	C	D	E	F	G	Empty columns			
1	1	1	1	1	1	1	1	1	1	1	1
2	-1	1	-1	1	1	1	-1	-1	-1	1	-1
3	-1	-1	1	-1	1	1	1	-1	-1	-1	1
4	1	-1	-1	1	-1	1	1	1	-1	-1	-1
5	-1	1	-1	-1	1	-1	1	1	1	-1	-1
6	-1	-1	1	-1	-1	1	-1	1	1	1	-1
7	-1	-1	-1	1	-1	-1	1	-1	1	1	1
8	1	-1	-1	-1	1	-1	-1	1	-1	1	1
9	1	1	-1	-1	-1	1	-1	-1	1	-1	1
10	1	1	1	-1	-1	-1	1	-1	-1	1	-1
11	-1	1	1	1	-1	-1	-1	1	-1	-1	1
12	1	-1	1	1	1	-1	-1	-1	1	-1	-1

9.3   Since robust design is about preventing variation to propagate, we want to find values of the control variables x so that $(g'_z)^2$ in

$$\sigma_y^2 \approx (g'_z)^2 \sigma_z^2$$

is small. The quantity $(g'_z)^2$ takes its minimum when

$$\frac{\partial (g'_z)^2}{\partial x} = 0.$$

Apply this to the equation in Example 9.2 and reflect on the result.

References

Fang KT, Li R, and Sudjianto A (2006) *Design and modeling for computer experiments.* Chapman & Hall/CRC.

Hjorth U (1994) *Computer intensive statistical methods. Validation, model selection, and bootstrap.* Chapman & Hall/CRC.

Sacks J, Schiller SB, and Welch WJ (1989a) Designs for computer experiments. *Technometrics*, **31**, 41–47.

Sacks J, Welch WJ, Mitchell TJ, and Wynn HP (1989b) Design and analysis of computer experiments (with discussion). *Statistica Sinica*, **4**, 409–435.

Santner TJ, Williams BJ, and Notz WI (2003) *The design and analysis of computer experiments.* Springer-Verlag.

10

Monte Carlo methods for robust design

10.1 Geometry variation

In many industries, such as cell phone and automotive, geometry variation plays an important role. There are primarily two reasons. One is to make it possible to manufacture. For a cell phone, for example, there are a large number of parts inside the space within the cell phone cover, and with a large variation of parts it may not even be possible to get the cover in place. Another aspect is the quality impression, that the product looks good. Take, for example, the side door of a car. If there is a large variation in geometry, the nominal gap between the door and the body side has to be large to ensure that there is a possibility of mounting. A large gap may look bad but is often acceptable as long as its size is the same along the entire door side. However, if it is 7 mm at the top end and 2 mm at the bottom end of the door, then it will certainly not look good and is not acceptable in the eyes of most customers.

Besides appearance and manufacturability, there is still another side of geometry variation, namely function. For example, in the manufacture of mechanical locks, such as door locks, it is obvious that a large geometry variation can endanger the function.

Consider once again the gap between the door and the body side. It will depend on a large number of components and manufacturing operations (i.e. both part-to-part variation and assembly variation). The geometry of the door itself as well as the car body, the hinges, and the operation to put the door in place are just some. This can cause a variation not just in the gap, which is often called fit in these applications, but

Statistical Robust Design: An Industrial Perspective, First Edition. Magnus Arnér.
© 2014 John Wiley & Sons, Ltd. Published 2014 by John Wiley & Sons, Ltd.
Companion website: www.wiley.com/go/robust

also in the difference in height between the door and the body, the so-called flush. Just as the fit at the top and bottom ends of the door can be different, so can the flush. In fact, there can be a variation in six different dimensions, namely transversal movements along the three axes of the coordinate system (x, y, and z directions) and rotation around all three of them.

From a computational point of view, geometry calculations may at first seem very easy. However, it does not require many components to be mounted together before some specialized software is required, even if only the nominal values are considered. If variation around the nominal is also considered and a calculation of the variation propagation in six dimensions of variation is needed, then calculations by hand are impossible. A possibility is then to make use of the methods of Chapter 9 and consider it just as any type of computer experiment. However, that is not needed, nor even desired. In contrast to cumbersome finite element calculations, the calculation time for geometry calculations is short. Thus, there is no need to consider response surfaces. Geometry has a great advantage, namely the possibility to visualize. With response surfaces, this opportunity will be lost. Thus, the problem is located somewhere in between the simple situation when hand calculations can be used and the complex situation when response surfaces are needed. One way that has turned out to be useful when studying variation of geometry is Monte Carlo simulation, which is the topic of this chapter.

Since this is a book about robust design, this is not a chapter about geometry variation in general, but first and foremost one of how to obtain robustness against the variation in the individual parts and in the assembly. Typically, it can be obtained by a good choice of connection points (sometimes referred to as reference or datum points) and types of connections between the parts. Thus, such positions are typical control factors. For example, if two metal sheets are mounted together, as in Figure 10.1, and the angle θ is the response (target value $0°$) and the variation in the positions of the holes the noise, then the response will be more robust if the holes are placed as far away from each other as possible.

Note 10.1   Monte Carlo simulation is not the only way to study variation propagation in geometry. A Taylor expansion is an alternative. The choice of analysis method depends on a number of things. Sometimes the computational burden is one, even though the calculations typically are very fast.

Note 10.2   A way to handle variation in positions of holes of the type shown in Figure 10.1 is to widen their diameters so that the hole diameter is considerably larger than the bolt diameter and make use of a washer or to have a larger diameter of the bolt head. Then it will still be possible to mount two metal sheets together as in Figure 1.3 or Figure 10.1. In addition, if the response is the mechanical tension shown in Figure 1.3 rather than the angle of misalignment, larger holes will make the design robust against the variation in positions of holes. However, there are some drawbacks. One is the angle θ shown in Figure 10.1. In industrial mass production, its variation will be even larger with larger holes (in contrast to manual production where this concept allows a fine tuning of the angle). Thus, we may decrease the

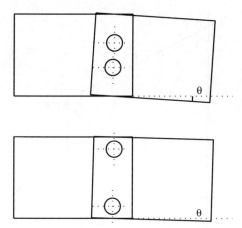

Figure 10.1 Two metal sheets are attached to each other. A small variation in the positions of the holes will cause a variation in angle θ.

sensitivity to part-to-part variation by adding an assembly variation with an equally bad end result. In addition, the washers will add another element in the manufacture and thus increase its complexity. This will make service more difficult since it needs to get fine tuned again after each time it has been taken apart.

10.1.1   Electronic circuits

Just as for geometry, the mathematics of electronic circuits is fairly easy. However, with increased complexity and size of circuits, hand calculation of variation propagation is not an option. Monte Carlo simulation turns out to be as equally useful as for geometry.

10.2   Geometry variation in two dimensions

To clarify the ideas of Monte Carlo simulation and how it can be used to obtain robustness of geometry, a two-dimensional example will be studied. Besides the Monte Carlo simulation itself, it will also be seen how to use ANOVA in this context. The results of ANOVA can be used to further reduce variation if robust design does not in itself give a small enough variance. ANOVA gives information about which tolerances to tighten if that is required or, vice versa, if the variation is unnecessary small and which tolerances to widen to reduce cost.

Example 10.1   Consider the framework of Figure 10.2. There are three bars A, B, and C. Two of them, A and C, are attached to a wall. The bar lengths as well as the angles are random variables. Assume that the position (x_2, y_2) is critical and needs to

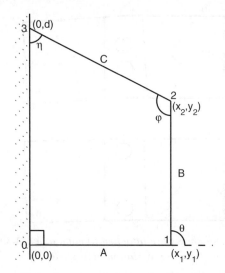

Figure 10.2   A conceptual sketch of the framework.

be close to its target value. The random variation of the lengths and angles will affect this vital position.

In this example, we will first select a robust 'concept' and will then optimize robustness with respect to this concept.

A connection between two bars or a bar and a wall (or any other type of connection) can be locked in one to six degrees. For a bolt and a nut connected to each other through a hole the connection is locked in six degrees. If two bars are connected just with a pin, they are free to rotate around one dimension. In our example with a two-dimensional framework, it is a rotation around the z axis. In a slot connection two dimensions of six are free, namely one rotation (here around z) and one transversal movement along x, or y, or a linear combination of them. To be able to mount the framework in the presence of random variation, all connections cannot be fixed in all six dimensions. Some of them need to be free to move, otherwise the geometry will be overdefined. On the other hand, if there are too many degrees of freedom, the geometry is underdefined.

For our framework, three choices will be explored, as shown in Figure 10.3 and Table 10.1.

For all the frameworks, the lengths of the bars A, B, and C are normally distributed with standard deviation $\sigma = 0.2$ and the angles are normal with standard deviation $\pi/100$ rad. For all the frameworks, the insertion points in the wall in connections 0 and 3 are nonrandom, and so is the angle in connection 0. These constants are

Point 0: $(0, 0)$

Point 3: $(0, d) = (0, 15)$

Angle 0: $\pi/2$ rad.

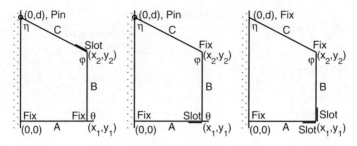

Figure 10.3   Sketches of the three implementations of the framework.

Framework 1

For framework 1, the random variables are

$$A \sim N(10, 0.2)$$
$$B \sim N(10, 0.2)$$
$$\theta \sim N\left(\frac{\pi}{2}, \frac{\pi}{100}\right).$$

Assume that bar C is 'long enough'. The angles η and ϕ will be given by the random variables and will not affect (x_2, y_2).

The position (x_2, y_2) is given by

$$x_2 = A + B\cos(\theta)$$
$$y_2 = B\sin(\theta).$$

A Monte Carlo simulation of $M = 10\,000$ runs gives the positions of Figure 10.4, where $\bar{x}_2 = 9.99$, $s_{x_2} = 0.37$, $\bar{y}_2 = 9.99$, and $s_{y_2} = 0.20$. The average distance $\sqrt{(x-10)^2 + (y-10)^2}$ to target is 0.37.

Table 10.1   The three frameworks of Example 10.1.

	Connection			
	1	2	3	4
Framework 1	Fix	Fix	Slot in bar C	Pin
Framework 2	Fix	Slot in bar A	Fix	Pin
Framework 3	Fix	Slot in bars A and B	Fix	Fix

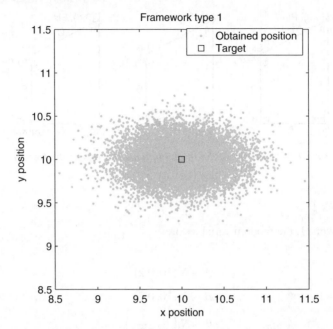

Figure 10.4   Simulation result of the position of (x_2, y_2) for framework 1.

A great possibility in the study of geometry variation is visualization. One way to visualize is to plot the simulated frameworks in one and the same graph. In order to do that, angle ϕ needs to be calculated,

$$\phi = \arctan\left(\frac{d - y_2}{x_2}\right) + \frac{\pi}{2}.$$

Since 10 000 frameworks in the same graph may make it impossible to discern anything at all, just the first 50 simulations are plotted in Figure 10.5.

Framework 2

For framework 2, the lengths of two bars and one angle need to be simulated,

$$B \sim N(10, 0.2)$$

$$C \sim N\left(\sqrt{(d - 10)^2 + 10^2}, 0.2\right) = N\left(\sqrt{125}, 0.2\right)$$

$$\phi \sim N\left(\arctan\left(\frac{d - 10}{10}\right) + \frac{\pi}{2}, \frac{\pi}{100}\right).$$

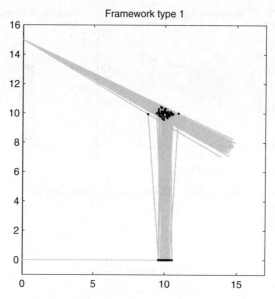

Figure 10.5   Simulation of framework 1.

Bar A will be 'long enough'. To calculate (x_2, y_2), define the distance E and the angles τ and v of Figure 10.6. The law of cosines gives

$$E = \sqrt{B^2 + C^2 - 2BC \cos(\phi)}$$

and Pythagoras' law

$$x_1 = \sqrt{E^2 - D^2}.$$

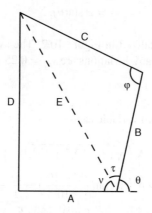

Figure 10.6   The angles in the calculation.

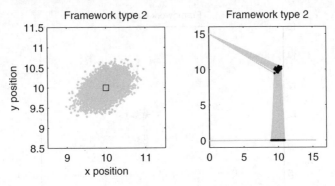

Figure 10.7   Simulation of framework 2.

Clearly,

$$v = \arcsin\left(\frac{D}{E}\right)$$

and the law of sines gives

$$\tau = \arcsin\left(\frac{C\sin(\phi)}{E}\right).$$

Thus,

$$\theta = \pi - \tau - v$$

and

$$x_2 = x_1 + B\cos(\theta)$$
$$y_2 = B\sin(\theta).$$

A simulation of this is visualized in Figure 10.7. The average values are $\bar{x}_2 = 9.99$ and $\bar{y}_2 = 9.99$ and the standard deviations are $s_{x_2} = 0.25$ and $s_{y_2} = 0.20$ respectively.

Framework 3

For framework 3, the random variables are

$$C \sim N\left(\sqrt{(d-10)^2 + 10^2}, 0.2\right)$$
$$\phi \sim N\left(\arctan\left(\frac{d-10}{10}\right) + \frac{\pi}{2}, \frac{\pi}{100}\right)$$
$$\eta \sim N\left(\arctan\left(\frac{10}{d-10}\right), \frac{\pi}{100}\right).$$

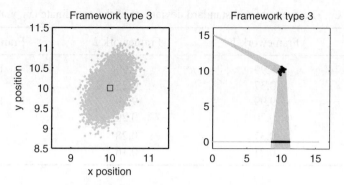

Figure 10.8   Simulation of framework 3.

Bars A and B will be 'long enough' and the angle θ is given by the random variables and will not affect (x_2, y_2).

More precisely,

$$x_2 = C\sin(\eta)$$
$$y_2 = D - C\cos(\eta)$$

give the position of the point of interest, (x_2, y_2) (Figure 10.8). To sketch the entire framework, note that

$$\theta = \eta + \phi - \pi/2.$$

Thus

$$B = \frac{y_2}{\sin(\theta)}$$

and

$$x_1 = x_2 - B\cos(\theta)$$

and the simulated frameworks can easily be calculated (Figure 10.8). The standard deviations become $s_{x_2} = 0.24$ and $s_{y_2} = 0.32$, and the means are $\bar{x}_2 = \bar{y}_2 = 10.00$.

The averages and standard deviations of the three frameworks are summarized in Table 10.2. In addition, the average and standard deviation of the distance to target,

$$dist = \sqrt{(x - 10)^2 + (y - 10)^2},$$

is also given in the table. Since framework 2 is the one with the smallest standard deviation and thus the most robust one, it will be studied further by using noise and control factors in a combined array.

Table 10.2 The averages and standard deviations of the coordinate (x_2, y_2).

	Framework 1	Framework 2	Framework 3
$\bar{x}$	9.99	9.99	10.00
s_x	0.37	0.25	0.24
$\bar{y}$	9.99	9.99	10.00
s_y	0.20	0.20	0.32
$\overline{dist}$	0.37	0.28	0.35
s_{dist}	0.21	0.16	0.20

A combined array

A typical control factor in geometry problems is positions of attachment points. In this example, the upper attachment point to the wall will be the control factor.

The nominal length of bar C is

$$\sqrt{(12 - 10)^2 + 10^2} \approx 10.2 \quad \text{if} \quad D = 12,$$

$$\sqrt{(17 - 10)^2 + 10^2} \approx 12.2 \quad \text{if} \quad D = 17.$$

The nominal value of the angle ϕ is

$$\arctan\left(\frac{12 - 10}{10}\right) + \frac{\pi}{2} \approx 1.77 \quad \text{if} \quad D = 12,$$

$$\arctan\left(\frac{17 - 10}{10}\right) + \frac{\pi}{2} \approx 2.18 \quad \text{if} \quad D = 17.$$

A combined array with the factor levels of Table 10.3 is set up. A 2^4 factorial with a centre point is run, or rather calculated (Table 10.4, with a coded level in Table 10.5). An experimental design of this type makes sense only if the standard deviation in bar length is the same no matter whether the nominal bar length is 10.2 or 12.2, and the standard deviation of the angle in the same way is independent of the nominal value.

A normal probability plot (Figure 10.10) reveals that there are some significant control-by-noise interactions for the position in the x direction, namely BD and CD.

Table 10.3 Factor and factor levels for the geometry example.

Factor	Type of factor	Low level	High level
ϕ	Noise	Nominal$-\pi/100$	Nominal$+\pi/100$
B	Noise	9.8	10.2
C	Noise	Nominal-0.2	Nominal$+0.2$
D	Control	12	17

Table 10.4   The experimental array.

	Noise		Control
ϕ	B	C	D
$\arctan(0.2) + 49\pi/100$	9.8	$\sqrt{104} - 0.2$	12
$\arctan(0.7) + 49\pi/100$	9.8	$\sqrt{149} - 0.2$	17
$\arctan(0.2) + 49\pi/100$	9.8	$\sqrt{104} + 0.2$	12
$\arctan(0.7) + 49\pi/100$	9.8	$\sqrt{149} + 0.2$	17
$\arctan(0.2) + 49\pi/100$	10.2	$\sqrt{104} - 0.2$	12
$\arctan(0.7) + 49\pi/100$	10.2	$\sqrt{149} - 0.2$	17
$\arctan(0.2) + 49\pi/100$	10.2	$\sqrt{104} + 0.2$	12
$\arctan(0.7) + 49\pi/100$	10.2	$\sqrt{149} + 0.2$	17
$\arctan(0.2) + 51\pi/100$	9.8	$\sqrt{104} - 0.2$	12
$\arctan(0.7) + 51\pi/100$	9.8	$\sqrt{149} - 0.2$	17
$\arctan(0.2) + 51\pi/100$	9.8	$\sqrt{104} + 0.2$	12
$\arctan(0.7) + 51\pi/100$	9.8	$\sqrt{149} + 0.2$	17
$\arctan(0.2) + 51\pi/100$	10.2	$\sqrt{104} - 0.2$	12
$\arctan(0.7) + 51\pi/100$	10.2	$\sqrt{149} - 0.2$	17
$\arctan(0.2) + 51\pi/100$	10.2	$\sqrt{104} + 0.2$	12
$\arctan(0.7) + 51\pi/100$	10.2	$\sqrt{149} + 0.2$	17
$\arctan(0.45) + \pi/2$	10.0	$\sqrt{120.25}$	14.5

A factor effect plot (Figure 10.9) shows that these interactions really may be useful to further increase robustness. However, there does not seem to be any way to increase the robustness of the y direction (Figures 10.10 and 10.11).

A graphical summary of the three originally explored framework implementations and the final, the optimal one, is given in Figure 10.12.

Analysis of variance

Analysis of variance is a way to understand the sources of variation in a better way. It can be used to identify the parameters for which the tolerances must be tightened in order to reduce the variation of the critical positions. To use ANOVA in this way is rather a second step in variance reduction, a way to further reduce the variation if robust design turns out to be insufficient or to identify the tolerances that can be widened if there are any. We will see how it can be applied to our problem with the frameworks. Only the linear terms of contribution are considered in this example.

Table 10.5   The experimental array and response (coded levels).

	Noise		Control	Response	
ϕ	B	C	D	x	y
−1	−1	−1	−1	9.75	9.78
−1	−1	−1	1	9.59	9.78
−1	−1	1	−1	10.16	9.79
−1	−1	1	1	10.10	9.79
−1	1	−1	−1	9.83	10.20
−1	1	−1	1	9.89	10.20
−1	1	1	−1	10.24	10.20
−1	1	1	1	10.38	10.20
1	−1	−1	−1	9.75	9.80
1	−1	−1	1	9.61	9.80
1	−1	1	−1	10.16	9.80
1	−1	1	1	10.10	9.80
1	1	−1	−1	9.83	10.19
1	1	−1	1	9.89	10.19
1	1	1	−1	10.24	10.19
1	1	1	1	10.36	10.18
0	0	0	0	10.00	10.00

The ANOVA calculation in a simple linear regression model is given by,

$$SS_{TOT} = SS_{Regression} + SS_{Error}$$
$$SS_{TOT} = \sum \left(y_i - \bar{y}\right)^2$$
$$SS_{Regression} = \sum \left(\hat{y}_i - \bar{y}\right)^2$$
$$SS_{Error} = \sum \left(y_i - \hat{y}_i\right)^2$$

where

$$\hat{y}_i = \hat{\beta}_0 + \hat{\beta}_1 x_i.$$

For a multiple regression model like the one we have here (for framework 2),

$$y = \hat{\beta}_0 + \hat{\beta}_1 x_B + \hat{\beta}_2 x_C + \hat{\beta}_3 x_\phi,$$

the calculations become somewhat more involved (in the following, something called 'constrained (type III) sums of squares' is used).

The ANOVA tables for framework 2 in its original form, with $D = 15$, are given in Tables 10.6 and 10.7. Obviously, the largest contributor to the variation in x is

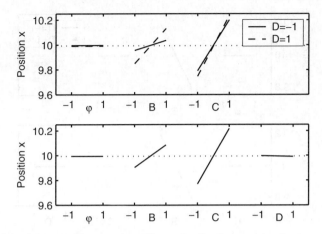

Figure 10.9   Main and interaction effects plot for the position in the x direction.

the bar length C and the largest contributor to the variation of y is the length B. The ANOVA tables for the optimal framework are given in Tables 10.8 to 10.9. Since the number of Monte Carlo runs is large, 10 000 to be accurate, even the smallest contribution is statistically significant even though it is of no practical relevance. The angle ϕ illustrates this; in all the tables it has a p value of zero so its contribution is statistically significant but it is totally irrelevant to use for further variance reduction. Thus, it is more rewarding to study the sum of squares than the p values. In that way it is easy to find the large contributors to the variation.

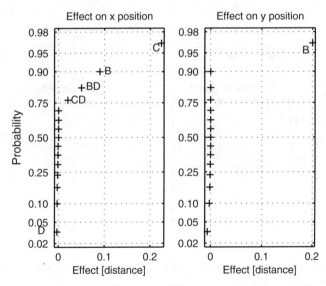

Figure 10.10   A normal probability plot of the estimated effects reveals some potentially useful control-by-noise interactions.

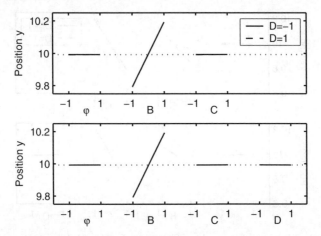

Figure 10.11   Main and interaction effects plot for the position in the y direction.

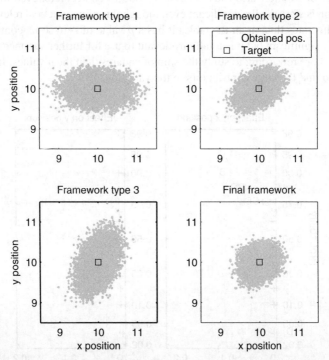

Figure 10.12   Graphical summary of the three original framework constellations and the final one.

Table 10.6   ANOVA for x for framework 2 with $D = 15$.

Source	SS	d.f.	MS	F	p
B	99.6	1	99.6	1 645 557	0
C	510.8	1	510.8	8 439 327	0
ϕ	0.003	1	0.003	57	0
Error	0.6	9996	0.0		
Total	697.9	9999			

Table 10.7   ANOVA for y for framework 2 with $D = 15$.

Source	SS	d.f.	MS	F	p
B	394.5	1	394	3 660 072	0.00
C	0.0	1	0	4	0.04
ϕ	0.007	1	0.007	60	0.00
Error	1.1	9996	0		
Total	395.6	9999			

Table 10.8   ANOVA for x for the final framework, $D = 12$.

Source	SS	d.f.	MS	F	p
B	15.8	1	15.8	916 836	0
C	421.5	1	421.5	24 382 694	0
ϕ	0.001	1	0.001	50	0
Error	0.2	9996	0.0		
Total	441.6	9999			

Table 10.9   ANOVA for y for the final framework, $D = 12$.

Source	SS	d.f.	MS	F	p
B	391	1	391	3 991 318	0
C	0	1	0	0.8	0.36
ϕ	0.004	1	0.004	42	0
Error	1.0	9996	0		
Total	393	9999			

Table 10.10   Crossed array with MC runs.

Control			Noise			Mean	Standard deviation
A	B	C	MC run 1	$\cdots$	MC run n		
-1	-1	-1	y_{11}	$\cdots$	y_{1n}	$\bar{y}_{1\cdot}$	s_1
$\vdots$	$\vdots$	$\vdots$	$\vdots$	$\ddots$	$\vdots$	$\vdots$	$\vdots$
1	1	1	y_{m1}	$\cdots$	y_{mn}	$\bar{y}_{m\cdot}$	s_m

10.3   Crossed arrays

In Example 10.1, there are only three noise and one control factors and the situation is fairly simple. In general, we cannot expect that to be the case. There may be many factors of both types. Especially the number of noise factors can be over whelming. Running a combined array as in Example 10.1 is not an option. It may be better to set up the inner array with control factors and in the outer array just have replicates of Monte Carlo runs, as in Table 10.10.

The approach can be used in a wide variety of applications. Geometry and electrical circuits are two good examples. However, we will take an example from another field, the deflection δ of an end loaded cantilever beam (Figure 10.13). The deflection is given by

$$\delta = \frac{FL^3}{3EI}$$

where

F = the applied force

L = beam length

E = elasticity modulus

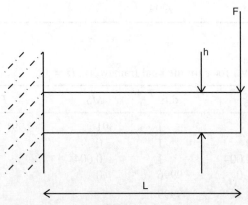

Figure 10.13   An end loaded cantilever beam.

Table 10.11   Results of the cantilever beam simulation (in nanometres).

| Control | | | Noise | | | | Standard |
L	b	h	Run 1	...	Run 1000	Mean	deviation
4.8	0.2	0.2	18 050	...	14 943	15 554	2584
4.8	0.2	0.4	2093	...	1856	1928	198
4.8	0.4	0.2	8735	...	7593	7755	1236
4.8	0.4	0.4	1013	...	943	961	88
5.6	0.2	0.2	28 777	...	23 619	24 701	4076
5.6	0.2	0.4	3337	...	2933	3062	310
5.6	0.4	0.2	13 926	...	12 002	12 315	1948
5.6	0.4	0.4	1615	...	1490	1527	137

and

$$I = \text{moment of inertia.}$$

If the beam section is rectangular with width b and height h, the moment of inertia is

$$I = \frac{bh^3}{12}.$$

Suppose that a small variation in deflection is desired. The deflection should be robust against random variation of the applied force, the width, height and length of the beam, and the elasticity modulus. The nominal values of beam dimensions, width, height and length, are the noise factors.

Suppose further that the noise follows normal distributions, more exactly

$$F \sim N(100, 3) \text{ [N]}$$
$$E \sim N(180, 5) \text{ [GPa]}$$

and the random variation around the nominal in length, width and height have standard deviations 0.05 m, 0.01 m and 0.01 m respectively. The levels of the beam dimensions along with the result are given in Table 10.11. An analysis shows that the height h of the beam is the most important contributor to the variation and should be selected at its low value to obtain a small variation of deflection.

11

Taguchi and his ideas on robust design

Robust design is so closely linked to its founder Genichi Taguchi that it is worth devoting an entire chapter to his contributions. His fundamental idea, to classify factors in the categories of noise and control, and then find a setting of the control factors making the design insensitive to the values taken by the noise, is splendid. Some other ideas of his leave room for improvement and some have given rise to extensive criticism. The Taguchi School has, unfortunately, not been very eager to incorporate ideas of others. A good example is combined arrays. When this book was written, more than 20 years after the first article about combined arrays, I have not seen a single example where the inner circle of the Taguchi School has used combined arrays. The reason can of course be that I have not looked deeply enough, but if it really had received the appreciation it deserves, I would not have had to look very deeply at all. However, I would also like to say that the Taguchi School has a point in almost all their ideas, however strange these ideas may seem. Take, for example, randomization. If an adherent of the Taguchi School is asked why they do not randomize their runs, the answer is that if there is a need for randomization, then there is something over which there is no control in laboratory testing and thus that the results cannot be trusted. There is certainly a point in that argument, even if it would not stop me from randomizing. It is like that with idea after idea. These ideas deserve an overview in this book.

11.1 History and origin

Taguchi was born in 1924 and was educated as a textile engineer. His home city had a tradition of kimono production, but this changed during the Second World War. He

---

Statistical Robust Design: An Industrial Perspective, First Edition. Magnus Arnér.
© 2014 John Wiley & Sons, Ltd. Published 2014 by John Wiley & Sons, Ltd.
Companion website: www.wiley.com/go/robust

came to work with something else. During his time in military service in the naviga-
tion unit in the fleet, he came into contact with the least squares method. It led him
into studies in statistics. His talent was observed by a professor at Tokyo University,
Motosaburo Masuyama. The professor sent him to industries as a consultant in design
of experiments. An example of his work during this period concerned a job at a candy
factory in the late 1940s. He planned an experiment where the hardness of candies
was studied and how this hardness depended on the temperature. The temperature
was a noise factor and the categorization of factors into two groups was already at
this time in his mind. However, it is in the 1950s that the ideas gained a more solid
foundation. Inner and outer arrays, for example, were applied to a problem of baking
tiles at the company Ina Seito in 1953. Actually, the example was mentioned earlier on
in this book (Example 3.5). During this time he visited the Indian Statistical Institute
with the prominent statistician Mahalanobis in 1955 and in 1960 Princeton University and Bell Laboratories in the United States and brought impressions and ideas
back home.

During the 1960s he started to use loss functions and the first form of SN ratios,
the one for digital systems. In the 1970s the family of SN ratios was extended with
the ratio for dynamic systems and somewhat later also the SN ratios for nondynamic
systems.

The breakthrough in the United States came in the wake of a television programme
in 1979, 'If Japan can, why can't we?' The programme was 1.5 hours long and was
broadcast at one of the major channels, NBC. Edward Deming, an American who
after the war often came to work with Japanese companies, was described as the
father of Japanese quality engineering. The programme became the start for a shift
from correcting problems afterwards to avoiding them to begin with. This opened
the way to the work of Taguchi, and he was primarily engaged by three American
companies, Xerox, Ford, and Bell Laboratories.

Two fairly late additions to the Taguchi School ideas are operating windows and
noise compounding that both were added after the breakthrough in the United States.
The first application of operating windows seems to be the one of the copy machine
(Example 6.3), that Phadke (1989) ascribes to Clausing and an internal memoran-
dum written at the company Xerox in 1980. Rather surprisingly, energy thinking is
also late, no matter how much it is associated with the most fundamental ideas of
Taguchi.

One experience that Taguchi had in his work is that requirements very often are
wrong. This is illustrated in Appendix A, where the loss function for a steering wheel
torque is estimated. It is not unusual for even the target value in a requirement not
to be correct, as the steering wheel example illustrates. It can even be various targets
depending on market or product variety. For the steering wheel torque it may be one
target for vehicles sold in Malaysia and another target for vehicles sold in Sweden
since the roads in these two countries are not equally crowned. Taguchi addressed
this by emphasizing the importance of a two-step approach, first identifying, a set of
control factors that can be used to increase the robustness and then another set that
can be used to tune to the target. The first set, the ones to improve robustness, will

immediately be applied to the design. The second set of control factors are kept to a later stage and used to tune to the target once it is known what this target is. This had some tradition to build upon in the Taguchi School since the two steps of first maximizing the *SN* ratio and then tuning to the target (or maximizing β) had been used for a long time. However, in the 1990s that two-step approach reached such a central position in the Taguchi School that it still retains it today. Earlier on the two steps were often taken in one instance by using a loss function. In the book by Phadke (1989) a lot of focus is on loss functions.

When robust design reached the United States and with some delay the rest of the world, it caught the eyes of the academic society. A number of eminent statisticians made important contributions and well aimed criticism. Combined arrays is one example. Unfortunately, the contributions were not appreciated by the Taguchi School, which may explain why the Taguchi School is still today considered as 'controversial'. Another reason can be the difference between business and research. Knowledge is shared for free in academics, but Taguchi had no intention to go to the universities and share his ideas for free. He was running a company.

11.2   The experimental arrays

When Taguchi started to work with the design of experiments, he had the foundations of Fisher and other pioneers to build upon. He added some elements and tweaked it to be better suited for product development and robust design. One of these elements is the categorization of factors. Some other elements are linked to the experimental array. Two aspects of this are worth mentioning. One is the idea to have two separate arrays, the inner and the outer, that are orthogonal to each other. The other is the nature of the inner array.

11.2.1   The nature of inner arrays

The choice of inner arrays in the Taguchi School is very limited, much too limited. There are primarily two reasons for this limitation. One is that control-by-control interactions are considered as uninteresting and should be nonexistent. The second reason is that if there are some control-by-control interactions anyway, their effects should be 'spread out' to minimize their impact on other estimates. Already in the early years of robust design, Taguchi made use of a type of array he called 'orthogonal array' for this purpose. This choice of name is a bit confusing. The different columns are indeed orthogonal, and so far the choice of name makes sense, but so are the columns of any 2^{k-p} experiment. The properties of the Taguchi orthogonal arrays resemble the properties that Plackett–Burman designs have with their complicated alias patterns. The different inner arrays of this Taguchi family of designs are called $L_{12}(2^{11})$, $L_{18}(2^1 \times 3^7)$, etc., and are discussed in more detail in Park (1996).

Let us take an example of the type of orthogonal array of this kind, an $L_{12}(2^{11})$ design. The design matrix is

$$X = \begin{pmatrix} -1 & -1 & -1 & -1 & -1 & -1 & -1 & -1 & -1 & -1 & -1 \\ -1 & -1 & -1 & -1 & -1 & 1 & 1 & 1 & 1 & 1 & 1 \\ -1 & -1 & 1 & 1 & 1 & -1 & -1 & -1 & 1 & 1 & 1 \\ -1 & 1 & -1 & 1 & 1 & -1 & 1 & 1 & -1 & -1 & 1 \\ -1 & 1 & 1 & -1 & 1 & 1 & -1 & 1 & -1 & 1 & -1 \\ -1 & 1 & 1 & 1 & -1 & 1 & 1 & -1 & 1 & -1 & -1 \\ 1 & -1 & 1 & 1 & -1 & -1 & 1 & 1 & -1 & 1 & -1 \\ 1 & -1 & 1 & -1 & 1 & 1 & 1 & -1 & -1 & -1 & 1 \\ 1 & -1 & -1 & 1 & 1 & 1 & -1 & 1 & 1 & -1 & -1 \\ 1 & 1 & 1 & -1 & -1 & -1 & -1 & 1 & 1 & -1 & 1 \\ 1 & 1 & -1 & 1 & -1 & 1 & -1 & -1 & -1 & 1 & 1 \\ 1 & 1 & -1 & -1 & 1 & -1 & 1 & -1 & 1 & 1 & -1 \end{pmatrix}. \tag{11.1}$$

The correlation matrix C of X is the identity matrix,

$$C = I,$$

so the columns are linearly independent and orthogonal. If the factors of the two first columns are A and B, then the A-by-B interaction has the design vector

$$x_{AB} = \begin{pmatrix} 1 & 1 & 1 & -1 & -1 & -1 & -1 & -1 & -1 & 1 & 1 & 1 \end{pmatrix}'.$$

This vector is not identical to any one of the columns of matrix X but is given by a linear combination of them,

$$x_{AB} = \frac{1}{3}X \begin{pmatrix} 0 & 0 & -1 & -1 & -1 & -1 & -1 & -1 & 1 & 1 & 1 \end{pmatrix}'.$$

The pattern of the other arrays, as $L_{18}(2^1 \times 3^7)$, resembles this but is even more complicated.

Designs of this kind are useful in screening experiments since it is an economical way to screen out the few vital factors to study further. However, in Taguchi's robust design, the experimental array is not used for screening but for the main experiment. Since the designs are highly reduced it is not possible to estimate interactions between control factors. According to Taguchi, it is not even necessary to estimate such interactions. It is just a matter of selecting the correct response variable. The 'correct response variable' is first and foremost something that is energy based. Since energy is additive, there is no need to estimate interactions. If some minor mistake is made in the selection of the response it will only have a small impact since possible interaction effects are 'spread out' in an even way due to the aliasing pattern of the arrays.

To understand the meaning of 'spread out', an example is now given. Assume that matrix X of Equation (11.1) is used as the inner array and that the true model of $\log(\sigma)$ is

$$\log(\sigma) = 0.7 + 0.3x_A + 0.3x_B + 0.3x_{AB}.$$

The interaction term will affect the estimates of the coefficients β_C to β_G, but to a third of its full value. Thus

$$E\left[\hat{\beta}_C\right] = -0.1.$$

The more factors there are, the larger the inner array will be and the impact of the interaction effects on the estimated main effects will be spread out over ever more estimates and have an ever less impact on each one of them. However, if the interaction is of a larger magnitude than the one of the main effects, the problem will remain. It is also worth noticing that the 'spreading out' just will reduce the problem if there really is a lack of interest of interactions and that there is a lack of interest no matter whether the interaction effects are small or large.

11.2.2   Interactions and energy thinking

As we have seen, Taguchi's view on interactions between control factors is that they should not exist, that their presence is simply a consequence of a bad choice of response. The argument is that it is the intended function that should be robust, and this is often an energy transformation. We have seen it on a large number of occasions in this book. Sometimes this is in terms of energy itself, but sometimes in terms of a quantity related to energy: an electrical motor whose intent is to transform electrical power into mechanical power or a force on a pedal that shall result in a pedal depression, thus transforming the applied energy into spring energy of the pedal. This energy thinking should guarantee that it is not necessary to take interaction effects into consideration. An example of a bad response variable is yield.

My own experience is that energy thinking helps. It seems as if useful conclusions are reached more often if the response is something that is energy related and the function to make robust is energy transformation. However, it does not mean that interactions between control factors should be disregarded. The trust I have in my own way to think in energy terms is far too limited for that. In addition, the arguments to disregard them are not convincing. If the principle of energy additivity is pushed to its limits, it ought to be valid for all types of factors, not just control factors. Since the possibility to accomplish robustness is based on the existence of control-by-noise interactions, this principle would undermine the very idea of robust design.

Another objection is that the response, seen from the view point of statistical analysis, rarely is y in itself, but, for example, $\log(\sigma)$ or one of Taguchi's *SN* ratios. Even if the terms are additive in y, they are not necessarily additive in *SN* ratios.

11.2.3   Crossing the arrays

When the advantages and disadvantages of inner and outer arrays were discussed in Chapter 5, we had not limited the choice of inner arrays to Taguchi's 'orthogonal arrays' only. We saw that inner and outer arrays are useful in many situations, even though combined arrays, which require less runs, is the default alternative.

The Taguchi way of using inner and outer arrays is to have an 'orthogonal array' for the control factors and a full factorial as the outer array. The alias pattern that this will give was discussed in Section 5.3.1. If we study this a bit further, we realize that it provides us with the possibility to estimate a number of interactions involving noise factors, such as noise-by-noise interactions and control-by-noise-by-noise interactions, but not control-by-control interactions. However, an engineer usually has more interest in control-by-control interactions than noise-by-noise interactions. The former can be used in product design, the latter cannot. Thus, it may be a better idea to have a full factorial as the inner array and a fractional factorial as the outer array. Besides the possibility of estimating control factor interaction effects, it will also use the data more efficiently. Since the observations for each row of the inner array are summarized in an average and a standard deviation (or possibly an *SN* ratio), some information is thrown away. If the inner array is of the type 2^{3-1} and the outer array 2^3, then the 32 observations are summarized into four mean values and four standard deviations. On the other hand, if the inner array is 2^3 and the outer array 2^{3-1}, then the 32 observations are summarized into eight mean values and the same number of standard deviations. Thus, the degrees of freedom is higher.

To summarize, using inner and outer arrays is often a good choice, but only if the candidate set for inner arrays is broader than Taguchi type arrays. Unfortunately, most commercial software do not provide this broader choice.

11.3   Signal-to-noise ratios

One of the ideas of the Taguchi School that has gained a solid grip in industrial practice but has been received with much less enthusiasm among statisticians is signal-to-noise ratios, for short *SN* ratios. Taguchi has taken the terminology from communication theory. The phrase is somewhat confusing since 'noise' does not mean 'sources of variation' in this case, but random variation in the response y.

There are several different *SN* ratios. The most common ones are the dynamic *SN* ratio

$$\eta = 10\log_{10}\left(\frac{\beta^2}{\sigma^2}\right)$$

and two different ratios for nominal-the-best (NTB)

$$\eta = 10\log_{10}\left(\frac{\mu^2}{\sigma^2}\right) \text{ (NTB1)}$$

$$\eta = 10\log_{10}\left(\frac{1}{\sigma^2}\right) \text{ (NTB2)}.$$

In Chapter 6 we studied situations for smaller-the-better (STB), larger-the-better (LTB), and operating windows. The SN ratios for these three situations are typically estimated with

$$\hat{\eta} = 10\log_{10}\left(\frac{1}{\frac{1}{n}\sum y_i^2}\right) \text{ (STB)} \tag{11.2}$$

$$\hat{\eta} = 10\log_{10}\left(\frac{1}{\frac{1}{n}\sum \frac{1}{y_i^2}}\right) \text{ (LTB)}$$

and

$$\hat{\eta} = 10\log_{10}\left(\frac{1}{\frac{1}{n}\sum y_i^2 \frac{1}{n}\sum \frac{1}{v_i^2}}\right) \text{ (operating windows)}$$

where v_i is defined in Section 6.4 as the highest value for which a failure is avoided. When SN ratios are described in words by the Taguchi School, expressions like 'the desired over the undesired' are used.

The last three of these SN ratios are given in terms of estimates rather than parameters. The fact is that the Taguchi School does not pay much attention to making a distinction between the parameter η and its estimate $\hat{\eta}$. Even if it makes sense to interpret Equation (11.2) as an estimate of

$$\eta = 10\log_{10}\left(\frac{1}{E[Y^2]}\right)$$

a number of questions of the properties of this estimate will arise. These questions will be even more complicated for LTB and operating windows.

There are several reasons for statisticians not to embrace SN ratios in an equally uncritical manner as industrial practitioners. A major one is that the standard deviation and the variance, the two most basic parameters for expressing random variation, are hidden in a complicated formula and not explicitly expressed. The probability distribution of the estimate is complicated, which makes it difficult to estimate confidence intervals and to understand which factors are statistically significant. The similarity to the coefficient of variation, σ/μ, is of course obvious to statisticians, but the fact that it is turned upside down is by the same community just considered as a way to complicate things. That $\log_{10}$ rather than the natural logarithm is used is also considered as a drawback, even though only a minor one.

It is common among statisticians to work with the logarithm of the response rather than the response itself. We have come across it on several occasions in this book. The reasons are usually to get an additive and linear model (rather than a multiplicative and nonlinear one) so that regression can be used, and in that way avoid models that give meaningless predictions such as negative lifetimes. Other reasons for taking the logarithm are to stabilize the variance so that it is roughly the same for all values of x in a regression model and to get a normal model. Typically, it is more common to consider $\log \sigma$ as a response rather than just σ. From this point of view, *SN* ratios should fit perfectly well into traditional statistical theory, but it is only partly true that it does.

Consider, for example, a set of lifetime data $(y_1, \ldots, y_n)$ that is assumed to be lognormal. In order to analyse it, the logarithm of each lifetime is taken and the average of the corresponding normal distribution is estimated with

$$\hat{\mu} = \frac{1}{n} \sum \log y_i. \tag{11.3}$$

Let us compare this with the *SN* ratio NTB1,

$$\eta = 10\log_{10}\left(\frac{\mu^2}{\sigma^2}\right) = 10\log_{10}\mu^2 - 10\log_{10}\sigma^2.$$

The first term can be estimated with

$$20\log_{10}\left(\frac{1}{n}\sum y_i\right). \tag{11.4}$$

Comparing Equations (11.3) and (11.4), we realize that the order of taking the logarithm and the average is reversed. This will change the statistical properties of the estimate. For example, $1/n \sum \log y_i$ is an unbiased estimate of $E[\log Y]$; $\log(\sum y_i/n)$ is not.

In many texts about robust design, no matter whether they are written by Taguchi or some statistician, there is a reluctance to use the *SN* ratios for STB and LTB: the statisticians due to the statistical properties (see Box, 1988), Taguchi since these two *SN* ratios take the focus away from the intended function of the product. For STB, the focus is on failure mode avoidance and for LTB on lifetime. The two ratios are just the last resort if nothing else works. Thus, the most interesting *SN* ratios are the dynamic and NTB. The *SN* ratio for these are partly there since higher average values are supposed to come hand in hand with a larger variance. The ratios σ/μ and σ/β, or rather the corresponding *SN* ratios, are variance stabilizing transformations. A variance stabilizing transformation is taken in order to have the same variance independently of the response, which facilitates statistical inferences considerably. In traditional statistics, variance stabilizing transformations are taken on the observation itself. The most common one is $\log y$, which is a special case in the Box–Cox class of transformations,

$$g_\lambda(y) = \begin{cases} \frac{y^\lambda - 1}{\lambda} & \text{if } \lambda \neq 0 \\ \log(y) & \text{if } \lambda = 0. \end{cases}$$

The *SN* ratios apply the transformation on the summary statistics s, $\hat{\mu}$, and $\hat{\beta}$ rather than on each one of the original observations. Box (1988) has a good and insightful discussion on this topic.

11.4   Some other ideas

11.4.1   Randomization

When I worked in the automotive industry, I had a colleague in the noise and vibration laboratory who was a frequent user of DOE. One day he investigated the level of audible noise in the passenger compartment. The experiment was a simple two-level factorial array, a 2^3 array to be more specific. The experiment was not randomized, and one factor, say factor C, was first run on its low level four times and then on its high level four times. After the first four runs of the experiment, another group of engineers passed by and asked if they could check something in the rear seat. Our friend from noise and vibration had nothing against this, so he took a short break and came back when the visitors had left in order to perform his last four runs.

The result of the data analysis from the experiment confused him. Factor C had by far the largest effect. It was in line with his expectations, but still a surprise. The effect was in the opposite direction from what his engineering experience and common sense told him. He had to investigate why, and realized that the other group had opened the ski lid, the lid that is found in the centre of the rear seat, thus causing an opening between the passenger compartment and the luggage compartment. This changed the acoustics substantially. What our friend had done was just an analysis of the effect of opening the ski lid. From that day on, he always randomized his tests to avoid issues of a similar kind

Just as the noise and vibration engineer, even if he had to learn his lesson the tough way, we need to randomize the run order of our experiments in order to avoid our conclusions to be affected by an unknown variation of uninteresting and unforeseen variables. It is recommended by every textbook in DOE. However, Taguchi did not randomize, and there is a rationale for his choice. If we need to randomize, then there is something in the experiment that we do not have control over, he argued, and we can then not trust our conclusions and believe that they will work in real action.

There is certainly a point in Taguchi's statement. Far too many experiments end up in useless conclusions since something has varied in an unforeseen way during the experiment. I have seen many such experiments, too many. In several of these cases, the problem could have been avoided by using a better measurement system, better communication between analyst and operator, or better control of the environmental conditions or test specimen. Randomization should not be a substitute for this. So far, Taguchi is perfectly right, but randomization can be seen as some kind of insurance; there will always be unforeseen variation and unexpected events taking place during the test. Therefore, in contrast to Taguchi, I believe in the value of randomization.

Somewhat contradictory to this is Taguchi's position on confirmation runs. He advocated the use of them. Once a robust design experiment is run and the design is optimized there should be yet one more experiment performed in this optimal setting (Chapter 6). Especially if the heavily reduced Taguchi type of inner arrays are used, such a confirmation run is of great importance since there may be unmodelled control-by-control interactions. However, the contradiction is that Taguchi also suggested that one of the rows of the inner array that already has been run (called the baseline) should be run again. The reason is that it often takes a while from the moment of the original runs to the confirmation run, so the baseline is run again to make certain that there has not been a level shift in the response. This is a good idea, but it is difficult to see that this argument can be combined with Taguchi's argument on the lack of need to randomize. To me, the two arguments are contradicting.

11.4.2   Science versus engineering

A fundamental element in the choice of experimental design for robust design, the choice between combined arrays and crossed arrays, is the number of runs they require. Furthermore, the freedom to study any interaction that can be of interest is bigger for combined arrays. This freedom will also make it possible to reach a deeper physical understanding. It is simply possible to obtain knowledge about which control factor it is that cancels out the effect of a certain noise. It is not possible to reach this insight if crossed arrays are used. Crossed arrays will only give knowledge that a certain control factor can be used to cancel out the effect of noise in general, not information about a specific noise.

Taguchi did not consider this as a drawback. The argument is that scientists want to understand how things work, engineers just want to make it work. The detailed knowledge that can be obtained by using a combined array is thus, for the Taguchi School, not of interest for engineers. On this point, I am convinced that Taguchi is wrong. There is, not the least, a curiosity among engineers and they want to understand. The knowledge is actually also of importance for the business. Without knowledge about how things work there is no possibility to solve problems quickly, problems experienced only after the product launch, and no possibility to meet new challenges such as changes in the environment and usage conditions among the customers. Last but not least, there would be limited possibilities to build knowledge that can be transferred to the next generation of engineers in the company.

11.4.3   Line fitting for dynamic models

In all the examples of dynamic models in this book, there has been a linear relation between the signal and response, and in most cases it has passed through the origin.

Table 11.1   Fitting a regression line through the origin or not will give different conclusions concerning which design is the most robust one.

	Signal							
	1	1	2	2	3	3	4	4
	Noise							
Design	−1	1	−1	1	−1	1	−1	1
1	0.17	0.41	0.47	0.63	0.89	1.01	1.11	1.41
2	0.01	0.07	0.27	0.38	0.95	1.11	1.25	1.33

This is in line with Taguchi. Even though modern Taguchi methods not are limited to such linear models, but also make use of other types of relations, even nonlinear, the regression line forced through the origin play a central role. The reason is energy. Zero energy in makes zero energy out. It was touched upon in Section 5.3 where signal factors in combined and crossed arrays were compared, but here we will look at it in more detail.

It may sometimes seem strange not to fit the model

$$y = \beta_0 + \beta_1 M + \varepsilon$$

but

$$y = \beta M + \varepsilon.$$

However, Taguchi has a point. Consider, for example, the data of Table 11.1 and Figure 11.1. Data as in design 2 are rather common in reality. It may be required that the added energy exceeds a certain value before any energy goes into the response. It may, for example, be some friction in the system that must be overcome. If the regression line is forced through the origin, the system is penalized for this behaviour. Thus, we are more likely to choose a design that does not have this threshold value to overcome.

The estimated SN ratios with the model

$$y = \theta_0 + \theta_1 M$$

are

$$10\log_{10}\left(\frac{\hat{\theta}_1}{s}\right)^2 = 15.9$$

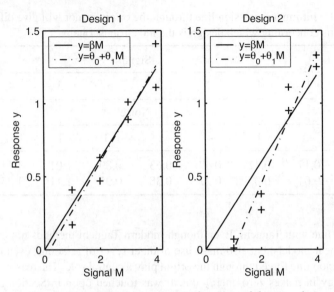

Figure 11.1   The response as a function of the signal.

for design 1 and

$$10\log_{10}\left(\frac{\hat{\theta}_1}{s}\right)^2 = 18.6$$

for design 2. For the model

$$y = \beta M \tag{11.5}$$

the estimates are 16.5 for design 1 and 10.8 for design 2. It is obvious that a free line fit will lead to undesired conclusions in this case if the *SN* ratio is used for design selection. However, it is also worth noticing that Equation (11.5) will be useless for forecasting the outcome given a certain signal level. Thus, it is possible that we should have different models for robust optimization, on the one hand, and for forecasting and analytical exploration, on the other hand.

11.4.4   An aspect on the noise

A question that almost every engineer asks themselves in robust design is whether or not they have missed any important noise factors in their experiments, noise that should have been considered as factors in the experiment. Sometimes this noise will vary anyway during the experiment, but in an uncontrolled and unknown manner. Then it is important to have a large enough number of replicates to increase the chance that the entire range of variation is exercised. Sometimes this noise does not vary by itself. If it does not and this is known to the experimenter, it is usually a better situation since in that case the experimenter has at least control of what is going on. However, if there are noise factors that are not deliberately changed in the experiment, since they are not in the experimental design and not varying randomly during the experiment, then we cannot know if the design is robust against this noise.

The Taguchi School has a very practical approach to this problem; if the design is robust against the most important noise factors, then it is also likely to be robust against other noise factors, even if they have not been explicitly tested. Such an argument is not based on science but probably on experience. Nevertheless, it seems likely that the argument holds in many situations.

11.4.5   Dynamic models

The focus that the Taguchi School has on dynamic models is strong. The book by Wu and Wu (2000) is an excellent example of this. The dynamic model is the default model. The statistical community, on the other hand, has not paid the same amount of interest to dynamic models. Furthermore, almost all articles on dynamic models deal with the existence of multiple targets as the reason for using a signal. It is certainly the case for a brake torque. A driver wants to brake harder sometimes and less hard at other times. However, very little attention has been paid to the fact that the cost of different control factor levels can be different. Take, for example, the set time for the

Table 11.2   A Taguchi type $L_{18}(2^1 \times 3^7)$ orthogonal array.

Run	A	B	C	D	E	F	G	H
1	−1	−1	−1	−1	−1	−1	−1	−1
2	−1	−1	0	0	0	0	0	0
3	−1	−1	1	1	1	1	1	1
4	−1	0	−1	−1	0	0	1	1
5	−1	0	0	0	1	1	−1	−1
6	−1	0	1	1	−1	−1	0	0
7	−1	1	−1	0	−1	1	0	1
8	−1	1	0	1	0	−1	1	−1
9	−1	1	1	−1	1	0	−1	0
10	1	−1	−1	1	1	0	0	−1
11	1	−1	0	−1	−1	1	1	0
12	1	−1	1	0	0	−1	−1	1
13	1	0	−1	0	1	−1	1	0
14	1	0	0	1	−1	0	−1	1
15	1	0	1	−1	0	1	0	−1
16	1	1	−1	1	0	1	−1	0
17	1	1	0	−1	1	−1	0	1
18	1	1	1	0	−1	0	1	−1

glue (Example 4.4). A longer time is more expensive. A useful approach is then to consider the time as a signal factor rather than a control factor, even though there is only one target value for the pull-off force. In that way, the expensive factor 'time' can be used more efficiently.

Dynamic models are one of Taguchi's brilliant ideas, but it has not yet received the recognition it deserves from the statistical community.

Exercises

11.1 Taguchi's $L_{18}(2^1 \times 3^7)$ array is given by Table 11.2.
   (a) Under the assumption that the factors are numerical, which linear combination of main effects will the interaction between factors A and C be confounded with?
   (b) One interaction is orthogonal to all main effects. Which one? How could this be used by the experimenter?

References

Box G (1988) Signal-to-noise ratios, performance criteria, and transformations. *Technometrics*, **30**, 1–17.

Park SH (1996) *Robust design and analysis for quality engineering*. Chapman & Hall/CRC.

Phadke MS (1989) *Quality engineering using robust design*. Prentice Hall.

Wu Y and Wu A (2000) *Taguchi methods for robust design*. ASME Press, Fairfield, New Jersey.

Appendix A

Loss functions

A.1 Why Americans do not buy American television sets

In the 1970s, the Japanese company Sony had one plant in Japan and one in the United States that manufactured identical television sets. The dealers in the United States who had units from both plants noted that customers preferred the ones manufactured in Japan. They wondered why and came to understand that the answer was to be found in the colour density. A comparison between the units from the United States and Japan along with the tolerance limits is given in Figure A.1. (The example comes from Phadke (1989), who in turn cites a Japanese newspaper, The *Asahi*.)

Does Figure A.1 lead us to the answer? The engineers were confused. In the US plant, all units were within the specification limits. Among the units from Japan, there were some that were actually outside the specification limits. In the view of the American engineers, customers should have preferred TV sets manufactured in the United States.

With this in mind, we must ask ourselves whether there is something fundamentally wrong with specification limits and their usage.

What the American engineers did not have in mind was the fact that there actually is a target value. The further away from the target, the larger the loss. For the colour density, the loss function might look as in Figure A.2. The figure shows that most Japanese units have a small loss since there is a concentration of units close to the target value. There are also many television sets from the US with a similar small loss, but there is also quite a large fraction of US units that have a loss of an intermediate size, with very few units from Japan in this category. The average loss of US television sets may very well be higher than those of the Japanese ones.

Statistical Robust Design: An Industrial Perspective, First Edition. Magnus Arnér.
© 2014 John Wiley & Sons, Ltd. Published 2014 by John Wiley & Sons, Ltd.
Companion website: www.wiley.com/go/robust

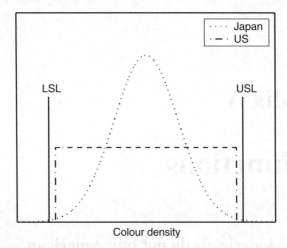

Figure A.1   Probability distributions and tolerance limits for the colour density. Phadke, Madhav S., Quality Engineering Using Robust Design, *1st Edition, © 1989. Reprinted by permission of Pearson Education, Inc., Upper Saddle River, NJ.*

In traditional tolerance thinking, it is tempting to consider loss as something that is zero inside the tolerance interval and one (or some other constant value) outside. However, does this really make sense? For the colour density, there is no step function for the loss. Some customers are dissatisfied if the colour density is just slightly off the target, a substantial proportion of them if it is considerably different from the

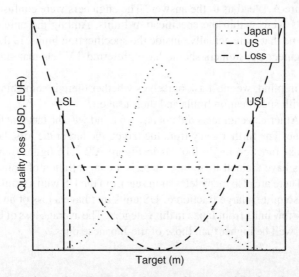

Figure A.2   Probability distributions, tolerance limits, and the loss function. Phadke, Madhav S., Quality Engineering Using Robust Design, *1st Edition, © 1989. Reprinted by permission of Pearson Education, Inc., Upper Saddle River, NJ.*

target. A loss function of the type in Figure A.2 makes more sense than traditional tolerance intervals.

A quadratic approximation of the loss

$$L(y) = K(y - m)^2$$

is useful, at least close to the target value. Such an approximation can be motivated using a Taylor expansion.

A.2   Taguchi's view on loss function

Even if it is possible to trace loss functions at least back to John von Neumann and the early works of game theory, it is primarily Taguchi who has spread the idea of loss functions in the industry. Thus, it might be in order to introduce some words on Taguchi's view on loss functions. Taguchi stated that the loss should be expressed in monetary units, as pointed out in Figure A.2, and that this should include all the cost that bad quality imposes on society. It includes dissatisfaction, waste of time, repair costs, etc. However, it also includes other costs, costs not limited to those of the producer and the customer. He even stated, maybe to exaggerate his point of view in order to make people understand it, that bad quality is worse than stealing. If a thief steals $100, the thief can use the money and it is not a total loss to society. That cannot be said about bad quality.

My personal view on loss functions is somewhat more pragmatic than that of Taguchi. To make them useful in practice, there are few occasions when it is worth considering any other costs than those of the producer and the customer. Other costs are only included if they come in a natural way. It happens quite often that the loss may not even have to be expressed in terms of money. It is often easier to measure the loss in terms of proportion of customers that are dissatisfied, which in practice is a perfectly good alternative. The consequence is that the distinction that sometimes is made between a product's 'quality loss function' and 'customer loss function' is far from always upheld.

A.3   The average loss and its elements

The expected value of the loss

$$E[L(Y)]$$

can be divided into its components,

$$
\begin{aligned}
E[L(Y)] &= E\left[K(Y - m)^2\right] = E\left[K\left(Y - \mu_y + \mu_y - m\right)^2\right] \\
&= KE\left[\left(Y - \mu_y\right)^2\right] + K\left(\mu_y - m\right)^2 \\
&= K\sigma_y^2 + K\left(\mu_y - m\right)^2.
\end{aligned}
$$

The second one of these two terms is a penalty for not aiming at the target, the first the penalty for having a variation around the mean.

Example A.1   Bergman and Klefsjö (2010) present an example about the torque that must be applied to the steering wheel of a car on a perfectly flat road (or possibly the angle of motion if no torque is applied). The data are not given explicitly there, but are simply in the form of diagrams without scales on the axis, so values given in this example are only the best guesses from these diagrams. The example is as follows.

The target value m of the torque applied to the steering wheel to drive straight ahead on a perfectly flat road should not be zero. The reason is that most roads are crowned to drain for water. It is the torque needed on normal roads, that is crowned roads, that should be zero. There is a problem in knowing exactly on which roads the drivers will drive and how crowned these roads are.

For a certain type of vehicle, this torque was noted for each and every vehicle that left the plant. The target value m of the torque was

$$m = 8.3.$$

It was also noted which drivers were dissatisfied with the torque of their vehicle. The data are given in Table A.1. The total number of vehicles in the sample is 40 475. The table should be read in such a way that among the 5425 drivers with vehicles with a torque of 6.5, 87 were dissatisfied. Thus, the proportion dissatisfied is $87/5425 = 0.016$.

Table A.1   The torque and adjustment data of Example A.1.

Torque	Number of vehicles	Number dissatisfied	Proportion dissatisfied
2.5	50	0	0.0000
3.5	200	6	0.0300
4.5	1875	48	0.0256
5.5	3150	58	0.0184
6.5	5425	87	0.0160
7.5	7325	72	0.0098
8.5	7650	61	0.0080
9.5	6450	31	0.0048
10.5	4200	14	0.0033
11.5	2275	7	0.0031
12.5	1100	4	0.0036
13.5	575	2	0.0035
14.5	150	1	0.0067
15.5	50	0	0.0000

To estimate the loss function, consider a regression model with the response

$$L = (0.030, 0.026, \ldots, 0.007)'$$

and the design matrix

$$X = \begin{pmatrix} 1 & 3.5 & 3.5^2 \\ 1 & 4.5 & 4.5^2 \\ \vdots & \vdots & \vdots \\ 1 & 14.5 & 14.5^2 \end{pmatrix}.$$

The estimated coefficients of the quadratic loss function are

$$\hat{\theta} = (X'X)^{-1}X'L = \begin{pmatrix} 0.0592 \\ -0.0095 \\ 0.0004 \end{pmatrix}.$$

The estimated regression model thus becomes

$$L = 0.0592 - 0.0095x + 0.0004x^2 = 0.0031 + 0.0004(x - 11.8085)^2$$

as sketched in Figure A.3. Obviously, the minimum loss is reached when the torque is 11.8, so this should be the target rather than 8.3. This observation gives input to a change of target.

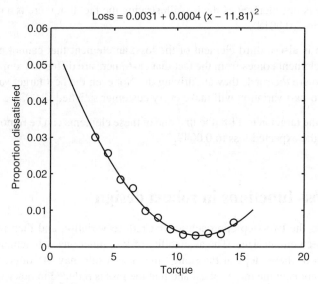

Figure A.3   The estimated loss function for the steering wheel torque. Reproduced by permission of Studentlitteratur AB.

Assume that the torque X is

$$X \sim N(\mu_x, \sigma_x).$$

The parameters μ_x and σ_x can be estimated with the mean and standard deviation of the sample,

$$\hat{\mu}_x = \bar{x} = 8.3,$$
$$\hat{\sigma}_x = s_x = 2.09.$$

The expected loss,

$$E\left[0.0031 + 0.0004(X - 11.8085)^2\right]$$
$$= 0.0031 + 0.0004E\left[(X - \mu_x)^2\right] + 0.0004(\mu_x - 11.8085)^2$$

can then be estimated by

$$0.0031 + 0.0004s_x^2 + 0.0004(\bar{x} - 11.8085)^2$$
$$= 0.0031 + 0.0018 + 0.0048 = 0.0097.$$

Obviously, the loss consists of three elements:

- There is one element of the loss illustrating the fact that the target does not reflect customer wants (0.0048).

- There is one element of the loss illustrating the fact that there is a part-to-part variation (0.0018).

- There is also a third element of the loss, an element that cannot be avoided. This element comes from the fact that customers are different from each other, and so are the roads they are driving on. Not even the best target value and no part-to-part variation will make every customer satisfied (0.0031).

By moving the target m to 11.8, the first one of these elements can be removed, which will reduce the expected loss to 0.0049.

A.4   Loss functions in robust design

In this book, the two-step approach to first reduce variation and then move to the target has been emphasized. The availability of loss functions will actually allow us to take both of these steps at the same time. The result may of course be that the design will not become more robust and that the loss is reduced in other ways. From a practical point of view this may seem irrelevant, and so it is. To the customer it is of no interest whatsoever if the loss has been reduced by a change in the target value or a reduction of the variation, and if it has been done by reducing the variance the customer is still not interested if this has been accomplished with robust design or in some other way.

Loss functions were first popularized by Taguchi for the purpose of robust design. Listening to the Taguchi School today, it is obvious that they have changed their view on the loss function and now advocate the two-step approach. There is a reason for this change of attitude, one of great importance. If the first step is to reduce the variation, a set of factors that can be used for this purpose is identified and their values fixed to levels minimizing the variation. This will give the possibility to work more freely with the remaining factors and use them to move to the target without affecting the variation. Another set of factors are identified that can be used to move the average without affecting the variation. By minimizing the loss function, this partitioning of factors cannot be accomplished. The consequence is that if during the course of the development it is realized that the target values are wrong, then the optimization work needs to start all over again. For the two-step approach, this is not the case. Requirements are often wrong in the beginning of the development process (and unfortunately also at the end of it). The steering wheel torque is one example. Using the same way of reasoning, a two-step approach would allow us to design several varieties of the same product, each variety with a different target but all of them with a small variation. In automotive engineering, for example, this is of great importance with many different carlines based on the same common platform.

Exercises

A.1 Consider the steering wheel torque in Example A.1.

(a) Calculate the average loss without using the estimated model. Compare with the results of Example A.1. If possible, divide this average loss into its components.

(b) Since there are fairly few vehicles with a steering wheel torque of 3.5, but many with 8.5, the latter should have a higher weight than the first one. Make use of weighted regression to estimate the loss function and compare with the one in Example A.1.

(c) Calculate confidence bands for the original loss function.

A.2 Lorenzen and Anderson (1993). The intermediate shaft steering column (Figure A.4) connects the steering wheel to the power steering motor. In the manufacture, the tube, which has a diameter slightly larger than the yoke, is slipped over the yoke and its end is crimped to get into the pockets of the yoke. If this connection becomes loose, an undesirable play is detected in the steering wheel. If the yoke and the tube are too tight, the assembly may not come apart as desired in a crash or for repairs. The part is assembled by slipping the tube over the yoke and crimping the end of the tube into pockets.

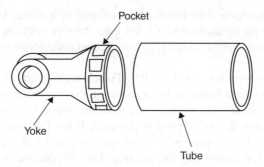

Figure A.4   The intermediate shaft steering column. Reproduced by permission of Taylor & Francis.

A torque to separate of 30 N m is considered ideal, with the loss function

$$L(y) = (y - 30)^2 .$$

An inner and outer array experiment with four control factors and two noise factors was set up. The control factors are

- *A*: pocket depth;
- *B*: yoke concentricity;
- *C*: tube length;
- *D*: power setting.

There were two noise factors, the play between the tube and yoke (*E*) and a variation in voltage, and thus in power, around the nominal value (*F*). The experiment resulted in the measurements given in Table A.2.

(a) Analyse the data in a two-step approach by first minimizing σ (or $\log \sigma$) and then move to the target.

(b) Analyse the data by minimizing the loss. Compare the conclusions with the one from the two-step approach. Which one makes more sense?

Table A.2   Torque data for the intermediate shaft steering column. Reproduced by permission of Taylor & Francis.

A	B	C	D	E F	−1 −1	−1 1	1 −1	1 1	Mean loss	Log loss
−1	−1	−1	−1		23.4	33.4	6.7	17.8	186.7	5.23
−1	−1	1	1		34.7	21.7	17.9	6.3	199.8	5.30
−1	1	−1	1		33.8	22.3	18.1	7.0	186.1	5.23
−1	1	1	−1		22.6	34.1	7.1	16.9	191.9	5.26
1	−1	−1	−1		26.9	17.0	42.6	30.7	84.5	4.44
1	−1	1	1		16.5	28.6	33.5	43.1	92.0	4.52
1	1	−1	1		14.9	28.5	33.0	42.6	99.5	4.60
1	1	1	−1		28.1	14.6	43.6	32.4	107.9	4.68

References

Bergman B and B Klefsjö (2010) *Quality. From customer needs to customer satisfaction*, 3rd edn. Studentlitteratur, Lund.

Lorenzen TJ and Anderson VL (1993) *Design of experiments. A no-name approach*. Marcel Dekker.

Phadke MS (1989) *Quality engineering using robust design*. Prentice Hall.

Appendix B

Data for chapter 2

Table B.1 Results from the measurements of strip position.
(see Example 2.1, where the factors and experiment are explained.)

	Control factors					Noise	Strip
Run	A	B	C	D	E	N	position
1	Low	Ball	U	Low	Low	Cold	−0.5
2	Low	Plain	U	Low	Low	Hot	−0.5
3	Low	Ball	V	Low	Low	Hot	−0.3
4	Low	Plain	V	Low	Low	Cold	0.1
5	Low	Ball	U	High	Low	Hot	−0.4
6	Low	Plain	U	High	Low	Cold	0.0
7	Low	Ball	V	High	Low	Cold	−0.6
8	Low	Plain	V	High	Low	Hot	−0.2
9	Low	Ball	U	Low	High	Hot	−0.3
10	Low	Plain	U	Low	High	Cold	−0.2
11	Low	Ball	V	Low	High	Cold	−0.2
12	Low	Plain	V	Low	High	Hot	−0.2
13	Low	Ball	U	High	High	Cold	−0.1
14	Low	Plain	U	High	High	Hot	0.1
15	Low	Ball	V	High	High	Hot	−0.2
16	Low	Plain	V	High	High	Cold	−0.5
17	High	Ball	U	Low	Low	Hot	−0.2
18	High	Plain	U	Low	Low	Cold	−0.4
19	High	Ball	V	Low	Low	Cold	0.2

(continued)

Statistical Robust Design: An Industrial Perspective, First Edition. Magnus Arnér.
© 2014 John Wiley & Sons, Ltd. Published 2014 by John Wiley & Sons, Ltd.
Companion website: www.wiley.com/go/robust

Table B.1   *(Continued)*

Run	A	B	C	D	E	Noise N	Strip position
			Control factors				
20	High	Plain	V	Low	Low	Hot	0.2
21	High	Ball	U	High	Low	Cold	0.0
22	High	Plain	U	High	Low	Hot	−0.5
23	High	Ball	V	High	Low	Hot	−0.8
24	High	Plain	V	High	Low	Cold	−0.6
25	High	Ball	U	Low	High	Cold	0.3
26	High	Plain	U	Low	High	Hot	−0.2
27	High	Ball	V	Low	High	Hot	−0.1
28	High	Plain	V	Low	High	Cold	−0.4
29	High	Ball	U	High	High	Hot	−0.4
30	High	Plain	U	High	High	Cold	−0.5
31	High	Ball	V	High	High	Cold	0.1
32	High	Plain	V	High	High	Hot	−0.6

Table B.2   The ten replicates for the measurements of the strip position
(Example 2.1). The order is the same as the one in Table B.1.

	1	2	3	4	5	6	7	8	9	10
					Replicate number					
1	−0.5	−0.7	−0.3	−0.4	−0.4	−0.4	−0.6	−0.4	−0.6	−0.5
2	−0.6	−0.7	−0.4	−0.5	−0.4	−0.5	−0.5	−0.6	−0.5	−0.6
3	0.0	−0.3	−0.5	−0.5	−0.4	−0.3	−0.2	−0.3	−0.1	0.0
4	0.1	0.1	0.0	0.0	0.2	0.2	0.2	0.2	0.1	0.1
5	−0.4	−0.6	−0.6	−0.5	−0.4	−0.5	−0.5	−0.2	−0.1	−0.7
6	0.0	−0.1	0.0	−0.1	0.0	0.0	0.1	0.0	0.1	0.0
7	−0.5	−0.5	−0.6	−0.7	−0.5	−0.6	−0.7	−0.8	−0.5	−0.7
8	0.0	0.0	−0.2	−0.2	−0.1	−0.2	−0.2	−0.3	−0.2	−0.4
9	−0.1	−0.1	−0.1	−0.1	0.0	−0.6	−0.6	−0.3	−0.1	−0.9
10	−0.3	−0.2	0.0	0.0	−0.2	−0.2	−0.2	−0.2	−0.2	−0.3
11	−0.1	−0.2	−0.3	−0.2	−0.1	−0.2	−0.2	−0.1	−0.2	−0.1
12	−0.1	−0.2	−0.2	−0.1	−0.2	−0.3	−0.3	−0.3	−0.2	−0.3
13	−0.2	−0.2	−0.2	−0.1	−0.2	−0.1	0.1	0.0	−0.1	0.2
14	0.0	0.0	0.2	0.2	0.1	0.1	−0.1	−0.1	0.1	0.1
15	−0.2	−0.3	−0.2	−0.3	−0.2	−0.2	−0.2	−0.3	−0.1	−0.2
16	−0.4	−0.6	−0.5	−0.6	−0.5	−0.5	−0.5	−0.4	−0.5	−0.4
17	0.2	0.0	0.2	0.1	−0.1	−0.1	−0.6	−0.5	−0.5	−0.5

(continued)

Table B.2   *(Continued)*

| | \multicolumn{10}{c}{Replicate number} |
	1	2	3	4	5	6	7	8	9	10
18	−0.5	−0.4	−0.4	−0.4	−0.4	−0.4	−0.4	−0.4	−0.3	−0.4
19	0.5	0.3	0.3	0.3	0.3	0.3	−0.1	−0.1	−0.1	−0.2
20	−0.4	−0.4	−0.2	−0.2	0.5	0.7	0.4	0.4	0.4	0.4
21	−0.5	−0.5	−0.1	0.0	0.1	0.1	0.1	0.3	0.2	0.1
22	−0.8	−0.8	−0.8	−0.7	−0.8	−0.7	−0.1	−0.2	0.1	0.1
23	−0.3	−0.6	−1.0	−1.0	−1.0	−1.1	−1.0	−0.9	−0.4	−0.4
24	−0.3	−0.4	−0.5	−0.4	−0.9	−0.9	−0.7	−0.9	−0.6	−0.6
25	0.3	0.3	0.3	0.3	0.3	0.4	0.3	0.5	0.3	0.2
26	0.2	0.2	−0.5	−0.4	−0.5	−0.4	−0.5	−0.5	−0.1	0.0
27	−0.7	−0.6	−0.6	−0.6	0.1	0.1	0.3	0.4	0.4	0.5
28	0.1	0.0	−0.3	−0.3	−0.6	−0.5	−0.5	−0.5	−0.5	−0.6
29	0.1	0.2	0.1	0.0	−0.1	0.0	−1.1	−1.0	−0.9	−0.9
30	−0.5	−0.5	−0.7	−0.7	−0.5	−0.6	−0.5	−0.4	−0.4	−0.5
31	0.2	0.2	0.1	0.1	0.1	0.2	0.2	0.2	0.0	0.1
32	0.0	0.1	−1.1	−1.3	−1.0	−1.2	−1.0	−0.9	0.4	0.4

Table B.3   The standard deviation for each set of replicates of the measurements of the strip position (Example 2.1).

| | \multicolumn{5}{c}{Control factors} | Noise factor | Standard deviation | Log of s.d. |
	A	B	C	D	E	N	s	log(s)
1	Low	Ball	U	Low	Low	Low	0.12	−2.10
2	Low	Plain	U	Low	Low	High	0.09	−2.36
3	Low	Ball	V	Low	Low	High	0.18	−1.69
4	Low	Plain	V	Low	Low	Low	0.08	−2.54
5	Low	Ball	U	High	Low	High	0.18	−1.69
6	Low	Plain	U	High	Low	Low	0.07	−2.71
7	Low	Ball	V	High	Low	Low	0.11	−2.21
8	Low	Plain	V	High	Low	High	0.12	−2.10
9	Low	Ball	U	Low	High	High	0.30	−1.19
10	Low	Plain	U	Low	High	Low	0.10	−2.27
11	Low	Ball	V	Low	High	Low	0.07	−2.70
12	Low	Plain	V	Low	High	High	0.08	−2.54
13	Low	Ball	U	High	High	Low	0.14	−1.97
14	Low	Plain	U	High	High	High	0.11	−2.23

(continued)

Table B.3   *(Continued)*

		Control factors				Noise factor	Standard deviation	Log of s.d.
	A	B	C	D	E	N	s	log(s)
15	Low	Ball	V	High	High	High	0.06	−2.76
16	Low	Plain	V	High	High	Low	0.07	−2.61
17	High	Ball	U	Low	Low	High	0.32	−1.15
18	High	Plain	U	Low	Low	Low	0.05	−3.05
19	High	Ball	V	Low	Low	Low	0.25	−1.40
20	High	Plain	V	Low	Low	High	0.41	−0.89
21	High	Ball	U	High	Low	Low	0.27	−1.29
22	High	Plain	U	High	Low	High	0.39	−0.93
23	High	Ball	V	High	Low	High	0.31	−1.17
24	High	Plain	V	High	Low	Low	0.23	−1.49
25	High	Ball	U	Low	High	Low	0.08	−2.54
26	High	Plain	U	Low	High	High	0.30	−1.22
27	High	Ball	V	Low	High	High	0.49	−0.70
28	High	Plain	V	Low	High	Low	0.25	−1.41
29	High	Ball	U	High	High	High	0.54	−0.62
30	High	Plain	U	High	High	Low	0.11	−2.24
31	High	Ball	V	High	High	Low	0.07	−2.66
32	High	Plain	V	High	High	High	0.69	−0.36

Appendix C

Data for chapter 5

Statistical Robust Design: An Industrial Perspective, First Edition. Magnus Arnér.
© 2014 John Wiley & Sons, Ltd. Published 2014 by John Wiley & Sons, Ltd.
Companion website: www.wiley.com/go/robust

Table C.1 Epitaxial layer thickness (example 5.1). Data taken from Wu and Hamada (2000) that cite Kackar and Shoemaker (1986). This material is reproduced with permission of John Wiley & Sons, Inc.

Control factors								Noise factors							
								L = Bottom				L = Top			
A	B	C	D	E	F	G	H	M = 1	M = 2	M = 3	M = 4	M = 1	M = 2	M = 3	M = 4
−1	−1	−1	1	−1	−1	−1	−1	14.29	14.19	14.27	14.19	15.32	15.43	15.27	15.41
−1	−1	−1	1	−1	1	−1	1	14.80	14.72	14.70	14.76	14.93	14.90	14.92	15.13
−1	−1	1	−1	−1	1	1	1	13.88	13.92	13.85	14.08	14.01	13.94	14.21	14.08
−1	−1	1	−1	1	−1	−1	−1	13.41	13.48	13.59	13.52	14.24	14.26	14.39	14.37
−1	1	−1	−1	−1	1	−1	−1	14.17	14.04	14.14	14.08	14.15	14.17	14.15	14.28
−1	1	−1	−1	1	−1	1	−1	13.25	13.33	13.19	13.44	14.22	14.30	14.27	14.41
−1	1	1	1	−1	−1	1	−1	14.06	14.09	14.18	14.05	15.30	15.52	15.42	15.21
−1	1	1	1	1	1	−1	−1	14.31	14.41	14.68	14.58	15.01	15.06	15.57	15.47
1	−1	−1	1	−1	−1	1	1	13.73	13.29	12.65	13.27	14.91	14.80	14.19	14.63
1	−1	−1	1	1	1	−1	1	13.90	14.56	14.45	13.71	13.75	14.32	14.22	13.82
1	−1	1	−1	−1	1	1	−1	14.22	14.40	15.28	15.04	14.19	14.43	15.55	15.22
1	−1	1	−1	1	−1	1	1	13.52	13.58	14.28	13.84	14.56	14.47	15.23	15.11
1	1	−1	1	1	−1	−1	1	14.53	14.25	14.67	15.28	14.74	14.18	14.97	15.55
1	1	−1	1	1	1	1	−1	14.57	14.03	13.71	14.64	15.87	15.22	14.97	16.00
1	1	1	−1	−1	−1	−1	1	12.90	12.71	13.15	13.89	14.25	13.84	14.13	15.17
1	1	1	−1	1	1	1	1	13.95	14.08	14.11	13.60	13.81	14.07	14.43	13.69

Table C.2 The measurement results of the heat exchanger experiment, part 1 (example 5.2).

Control			Noise		Signal	Response replicate			Standard
A	B	C	D	E	M	1	2	3	deviation
−1	−1	−1	−1	−1	−1	5.9	6.7	6.1	0.44
−1	−1	−1	1	1	−1	5.4	5.3	5.4	0.06
−1	−1	1	−1	1	−1	4.9	4.6	4.5	0.20
−1	−1	1	1	−1	−1	6.4	6.9	6.8	0.25
−1	1	−1	−1	1	−1	4.7	4.7	4.7	0.03
−1	1	−1	1	−1	−1	6.9	7.7	7.4	0.41
−1	1	1	−1	−1	−1	6.9	6.5	6.2	0.32
−1	1	1	1	1	−1	5.8	5.6	5.5	0.15
1	−1	−1	−1	1	−1	5.3	5.3	5.4	0.07
1	−1	−1	1	−1	−1	6.3	6.3	6.2	0.07
1	−1	1	−1	−1	−1	7.0	7.2	7.5	0.24
1	−1	1	1	1	−1	4.9	4.3	4.3	0.33
1	1	−1	−1	−1	−1	7.7	7.8	7.6	0.13
1	1	−1	1	1	−1	5.1	5.2	4.8	0.18
1	1	1	−1	1	−1	5.6	5.9	6.2	0.32
1	1	1	1	−1	−1	7.0	6.3	6.8	0.35
−1	−1	−1	−1	−1	0	10.0	7.1	6.0	2.06
−1	−1	−1	1	1	0	7.2	7.2	7.7	0.28
−1	−1	1	−1	1	0	4.9	6.5	7.5	1.31
−1	−1	1	1	−1	0	8.6	8.5	9.0	0.24
−1	1	−1	−1	1	0	5.4	6.6	6.8	0.78
−1	1	−1	1	−1	0	9.0	7.4	8.1	0.80
−1	1	1	−1	−1	0	6.5	8.4	7.4	0.92
−1	1	1	1	1	0	8.1	7.3	8.6	0.62
1	−1	−1	−1	1	0	8.6	8.2	8.6	0.23
1	−1	−1	1	−1	0	9.5	8.9	8.2	0.66
1	−1	1	−1	−1	0	10.2	9.3	9.8	0.45
1	−1	1	1	1	0	6.7	7.3	7.3	0.36
1	1	−1	−1	−1	0	10.1	9.6	9.6	0.28
1	1	−1	1	1	0	8.2	7.5	7.1	0.54
1	1	1	−1	1	0	7.8	8.2	8.6	0.41
1	1	1	1	−1	0	9.2	8.5	7.9	0.61

Table C.3   The measurement results of the heat exchanger experiment, part 2.

| Control | | | Noise | | Signal | Response replicate | | | Standard |
A	B	C	D	E	M	1	2	3	deviation
−1	−1	−1	−1	1	1	10.6	9.5	5.9	2.47
−1	−1	−1	1	−1	1	6.3	11.3	12.9	3.48
−1	−1	1	−1	−1	1	6.1	6.9	7.9	0.89
−1	−1	1	1	1	1	11.4	8.8	15.1	3.17
−1	1	−1	−1	−1	1	8.5	10.0	8.1	1.01
−1	1	−1	1	1	1	11.4	10.4	10.7	0.50
−1	1	1	−1	1	1	12.3	12.7	11.4	0.64
−1	1	1	1	−1	1	10.3	5.9	9.0	2.25
1	−1	−1	−1	−1	1	9.4	9.1	9.7	0.35
1	−1	−1	1	1	1	11.7	9.6	12.6	1.52
1	−1	1	−1	1	1	11.2	14.4	10.3	2.15
1	−1	1	1	−1	1	8.5	10.2	10.7	1.15
1	1	−1	−1	1	1	14.8	11.4	11.3	2.00
1	1	−1	1	−1	1	10.3	10.2	11.2	0.54
1	1	1	−1	−1	1	9.7	10.5	12.1	1.22
1	1	1	1	1	1	12.0	12.5	10.0	1.30

Table C.4   The results of the fan data experiment (example 5.3).

		Outer array (noise and signal factors)														
1		1.4	3.5	5.0	13.9	5.8	7.8	9.9	11.1	19.7	21.6	24.1	30.5	44.9	45.7	46.2	46.6
2		0.0	2.8	7.5	7.9	8.5	13.9	15.2	16.8	20.5	26.0	28.3	27.6	43.3	45.5	52.0	55.9
3		6.3	6.2	7.5	9.0	8.7	12.7	17.5	16.9	24.2	25.5	24.5	28.9	47.0	48.5	51.5	52.7
4	I	4.1	8.4	8.0	12.0	3.9	8.3	12.6	13.8	21.9	21.9	25.4	24.4	49.8	49.1	50.2	55.4
5	n	7.3	6.0	10.1	10.4	6.6	9.2	9.5	11.4	18.0	22.0	22.0	22.1	44.4	45.6	45.7	48.3
6	n	6.7	8.1	9.2	11.0	8.4	9.2	14.7	14.8	21.5	24.1	27.9	27.5	46.4	50.6	50.7	57.9
7	e	1.9	6.0	7.2	10.3	13.4	14.7	15.9	16.9	26.6	27.7	28.3	31.5	53.7	54.7	54.6	57.4
8	r	5.0	5.5	6.3	7.4	11.1	13.3	13.1	14.7	24.9	23.3	26.0	25.8	49.6	49.8	49.3	50.3
9		2.9	3.3	3.9	4.4	11.1	9.1	14.8	17.0	21.1	22.0	21.7	23.4	39.2	42.9	43.8	45.3
10	a	4.1	3.8	7.1	7.3	12.5	12.1	13.9	13.5	22.5	25.5	26.9	30.0	50.5	54.3	55.3	57.3
11	r	1.5	2.8	10.3	13.1	0.0	1.3	2.9	13.0	12.2	14.2	13.6	17.6	29.3	28.7	32.6	43.6
12	r	0.0	4.6	3.5	13.7	12.8	16.5	18.3	18.2	22.3	26.1	26.9	32.8	34.7	45.0	50.7	59.4
13	r	2.7	4.7	5.4	13.4	10.8	11.5	13.7	20.5	26.3	30.4	31.7	33.2	52.2	55.2	58.2	59.3
14	a	3.9	5.4	7.8	8.2	10.5	10.8	11.8	12.9	20.8	21.8	22.6	23.0	46.0	47.3	47.3	48.6
15	y	0.0	0.0	4.9	3.4	0.6	6.8	14.1	14.3	17.8	17.4	23.4	28.3	30.1	32.7	32.1	36.0
16		6.5	6.4	8.2	8.3	11.9	11.6	14.8	14.4	25.2	25.0	26.3	28.7	52.6	58.2	58.3	58.0
17		0.8	4.7	6.0	7.7	7.8	10.8	13.5	21.4	18.9	21.4	24.0	26.3	46.0	46.0	48.4	48.7
18		0.0	2.4	8.0	8.7	11.0	11.2	13.1	13.4	15.7	16.8	17.1	19.6	36.9	46.0	45.3	50.8

Table C.5 The hammock data of Exercise 5.2.

				Noise factors													
				Package rotation : 0°						Package rotation : 5°							
Control factors				Package 1			Package 2			Package 1			Package 2				
A	B	C	D	Pos 1	Pos 2	Pos 3	Pos 1	Pos 2	Pos 3	Pos 1	Pos 2	Pos 3	Pos 1	Pos 2	Pos 3		
−1	−1	−1	−1	0.3	0.1	0.4	0.4	0.0	0.4	1.4	1.4	1.6	1.1	1.2	0.9		
1	−1	−1	1	−0.2	−0.8	−0.3	−0.8	−0.6	−1.1	1.5	1.8	1.7	1.7	1.5	1.4		
−1	1	−1	1	−0.3	−0.1	−0.1	−0.8	0.0	−0.5	1.7	1.6	1.5	1.5	1.2	1.8		
1	1	−1	−1	−1.1	−1.2	−0.8	−1.3	−1.0	−1.0	1.9	1.9	2.1	2.2	1.9	1.5		
−1	−1	1	1	0.6	0.4	0.6	0.4	0.5	0.3	1.2	1.4	1.3	1.3	1.4	1.1		
1	−1	1	−1	0.6	0.9	0.6	0.7	0.5	1.0	2.9	3.0	3.0	3.3	2.7	3.2		
−1	1	1	−1	0.2	0.2	−0.1	−0.3	0.4	−0.1	1.9	1.4	1.9	1.6	2.0	1.8		
1	1	1	1	0.7	0.8	0.3	0.4	0.2	0.4	3.0	3.4	3.1	3.3	3.6	3.1		

References

Kackar RN and Shoemaker AC (1986) Robust design: a cost-effective method for improving manufacturing processes. *AT & T Technical Journal*, 65, 39–50.

Wu J and Hamada M (2000) *Experiments, Planning, analysis, and parameter design optimization*. John Wiley & Sons, Inc.

Appendix D

Data for chapter 6

Table D.1 The experimental array of the wave soldering experiment. From Peace (1993).

A	B	C	D	E	F	G	H	I	J	K	L	M	N	O
−1	−1	−1	−1	−1	−1	−1	−1	−1	−1	−1	−1	−1	−1	−1
−1	−1	−1	−1	−1	−1	−1	1	1	1	1	1	1	1	1
−1	−1	−1	1	1	1	1	−1	−1	−1	−1	1	1	1	1
−1	−1	−1	1	1	1	1	1	1	1	1	−1	−1	−1	−1
−1	1	1	−1	−1	1	1	−1	−1	1	1	−1	−1	1	1
−1	1	1	−1	−1	1	1	1	1	−1	−1	1	1	−1	−1
−1	1	1	1	1	−1	−1	−1	−1	1	1	1	1	−1	−1
−1	1	1	1	1	−1	−1	1	1	−1	−1	−1	−1	1	1
1	−1	1	−1	1	−1	1	−1	1	−1	1	−1	1	−1	1
1	−1	1	−1	1	−1	1	1	−1	1	−1	1	−1	1	−1
1	−1	1	1	−1	1	−1	−1	1	−1	1	1	−1	1	−1
1	−1	1	1	−1	1	−1	1	−1	1	1	−1	1	−1	1
1	1	−1	−1	1	1	−1	−1	1	1	−1	−1	1	1	−1
1	1	−1	−1	1	1	−1	1	−1	−1	1	1	−1	−1	1
1	1	−1	1	−1	−1	1	−1	1	1	−1	1	−1	−1	1
1	1	−1	1	−1	−1	1	1	−1	−1	1	−1	1	1	−1

Statistical Robust Design: An Industrial Perspective, First Edition. Magnus Arnér.
© 2014 John Wiley & Sons, Ltd. Published 2014 by John Wiley & Sons, Ltd.
Companion website: www.wiley.com/go/robust

Table D.2   The results of the wave soldering experiment. From Peace (1993).

Row	Lower limit, y					Upper limit v				
1	247	245	242	245	240	253	260	265	265	250
2	235	232	230	232	230	231	235	238	238	240
3	229	223	220	225	220	273	280	290	280	275
4	234	230	235	233	228	222	230	235	230	228
5	242	235	234	235	230	228	235	234	235	230
6	242	230	238	234	237	252	260	258	264	257
7	237	234	235	230	232	248	255	245	260	252
8	238	235	236	235	230	234	240	246	235	240
9	241	240	235	240	235	270	275	285	270	275
10	230	225	222	215	215	215	225	222	215	225
11	224	220	215	212	212	261	268	265	262	272
12	231	230	228	228	226	225	230	238	228	236
13	239	235	235	230	235	235	235	235	240	245
14	239	235	238	235	230	235	235	238	245	250
15	223	220	215	218	218	245	255	265	238	258
16	222	220	215	224	218	255	260	265	244	268

Reference

Peace GS (1993) *Taguchi Methods*. Addison-Wesley.

Index

Statistical Robust Design: An Industrial Perspective, First Edition. Magnus Arnér.
© 2014 John Wiley & Sons, Ltd. Published 2014 by John Wiley & Sons, Ltd.
Companion website: www.wiley.com/go/robust